Systems of Commercial Turbofan Engines

Andreas Linke-Diesinger

Systems of Commercial Turbofan Engines

An Introduction to Systems Functions

Springer

Andreas Linke-Diesinger
21149 Hamburg
Germany
engine-systems@arcor.de

ISBN 978-3-540-73618-9 e-ISBN 978-3-540-73619-6

DOI 10.1007/978-3-540-73619-6

Library of Congress Control Number: 2007940818

Typesetting and production: LE-TEX Jelonek, Schmidt & Vöckler GbR, Leipzig, Germany
Cover design: Frido Steinen-Broo, eStudio Calamar, Spain

Printed on acid-free paper

9 8 7 6 5 4 3 2 1

springer.com

Preface

Many textbooks describing the function of a turbine engine are available on the market. These books describe the operation of the main components of an engine, its thermodynamic principles and the principle of thrust generation. They are available at any required level. For persons that first get in touch with the subject matter of turbine engines it is also important to understand the operation of the systems necessary for the operation of a turbine engine.

With my long-term experience as a technical training instructor the idea developed to write a description of the engine systems to provide this as an add-on to the many text books available describing the operation of the basic turbine engine. For a complete understanding of functions of the engine systems it is important not to describe the systems isolated from the aircraft. Systems of modern turbine engines are highly integrated into the aircraft system architecture. Thus the functional interface with the aircraft is shown were necessary.

It took some time to find a concept for this book that provides the information in a sequence that facilitates the understanding of the described functions. During the writing of this book I got some help from experts in their field for the optimization of the result. I wish to thank Johannes Lange and Dr. Hans Hansen for their expertise concerning the concept of this book; Sybille Schmidt and Wolfgang Lehmann for their useful linguistic suggestions.

I also mustn't forget to thank my family for their patience while I worked on this book for some months.

Hamburg, August 2007

Table of Contents

Abbreviations

AC	Alternating Current
ACC	Active Clearance Control
ACOC	Air-Cooled Oil Cooler
ADC	Air Data Computer
AFDX	Avionics Full Duplex Switched Ethernet
AIMS	Aircraft Information Management System
ALF	Aft Looking Forward
ARINC	Aeronautical Radio Inc.
ASC	Aircraft System Computer
ASTM	American Society for Testing and Materials
ATA	Air Transport Association of America
A/THR	Autothrust
AVM	Airborne Vibration Monitoring
CBP	Customer Bleed Pressure
CCS	Common Core System
CDP	Compressor Discharge Pressure
CIT	Compressor Inlet Temperature
CONT	Continuous
COTS	Commercial Off The Shelf
CS	Certification Standard
DAC	Dual Annular Combustor
DC	Direct Current
DEU	Display Electronic Unit
DMC	Display Management Computer
EASA	European Aviation Safety Agency
ECAM	Electronic Centralized Aircraft Monitor
ECM	Engine Condition Monitoring
ECU	Electronic Control Unit
EEC	Electronic Engine Control
EFIS	Elecronic Flight Instrument System
EHSV	Electrohydraulic Servo Valve
EIA	Electronic Industries Alliance

EICAS	Electronic Indication and Crew Alerting System
EIU	Engine Interface Unit
EIVMU	Engine Interface and Vibration Monitoring Unit
EPR	Engine Pressure Ratio
EUROCAE	European Organisation for Civil Aviation Equipment
E/WD	Engine/Warning Display
FADEC	Full Authority Digital Engine Control
FCU	Fuel Control Unit
FDRV	Fuel Diverter and Return Valve
FF	Fuel Flow
FIFO	First In First Out
FLA	Forward Looking Aft
FMC	Flight Management Computer
FMU	Fuel Metering Unit
FMV	Fuel Metering Valves
FOB	Fuel On Board
FRT	Flat Rate Temperature
ft	Feet
FWC	Flight Warning Computer
FWD	Forward
GRD	Ground
HMU	Hydromechanical Unit
HPC	High Pressure Compressor
HPT	High Pressure Turbine
IDG	Integrated Drive Generator
IEEE	Institute of Electrical and Electronics Engineers
IGN	Ignition
IMA	Integrated Modular Avionics
IPC	Intermediate Pressure Compressor
IPT	Intermediate Pressure Turbine
ISA	International Standard Atmosphere
LLP	Life Limited Part
LPC	Low Pressure Compressor
LPT	Low Pressure Turbine
LRM	Line Replaceable Modules
LRU	Line Replaceable Unit
LVDT	Linear Variable Differential Transformer
MCDU	Multipurpose Control and Display Unit
MEC	Main Engine Control

MEMS	Microelectromechanical Systems
Mn	Mach Number
N1, N2, N3	Engine Rotor Speeds
OAT	Outside Air Temperature
ODM	Oil Debris Monitor
PMC	Power Management Control
PPBU	Power Plant Build-Up
QEC	Qick Engine Change
REV	Reverse
RTD	Resistive Thermal Device
RTDCA	Radio Technical Commission for Aeronautics
RTOS	Real Time Operating Systems
RVDT	Rotary Variable Differential Transformer
SAL	System Address Label
SDAC	System Data Aquisition Computer
SOAP	Spectrographic Oil Analysis Program
TAPS	Twin Annular Premixing Swirler
TAT	Total Air Temperature
TBV	Transient Bleed Valve
TIA	Telecommunication Industry Association
TLA	Thrust Lever Angle
TOGA	Take-Off/Go Around
TRA	Thrust Lever Resolver Angle
TSFC	Thrust Specific Fuel Consumption
UART	Universal Asynchronous Receiver-Transmitter
VBV	Variable Bleed Valve
VSV	Variable Stator Vane

Company Names

Airbus	Airbus S.A.S., Blagnac Cedex, France
Boeing	Boeing Commercial Airplanes, Renton, Wash., USA
CFM	CFM International S.A., Melun, France
GEAE or GE	GE Aviation, Cincinnati, Ohio, USA
IAE	International Aero Engines, East Hartford, Ct., USA
LTT	Lufthansa Technical Training, Hamburg, Germany
Pratt & Whitney	Pratt & Whitney, East Hartford, Ct., USA
RR	Rolls-Royce International Ltd., London, UK

1 Introduction

1.1 Engine Systems in General

A turbine engine consists of its main components, which change the state of the gas flow in the sequence of the thermodynamic working cycle. The design of modern turbofan engines follows a modular concept. Thus a typical twin-spool turbofan engine, like the V2500-A5 shown in Fig. 1.1, is composed of the following main modules:

- Fan module
- Low pressure compressor module
- Core engine or gas generator
- Low pressure turbine module
- Accessory gearbox module

The core engine consists of the high pressure compressor, the combustion section and the high pressure turbine. This modular design of the engine mainly reflects maintenance aspects. During engine disassembly each

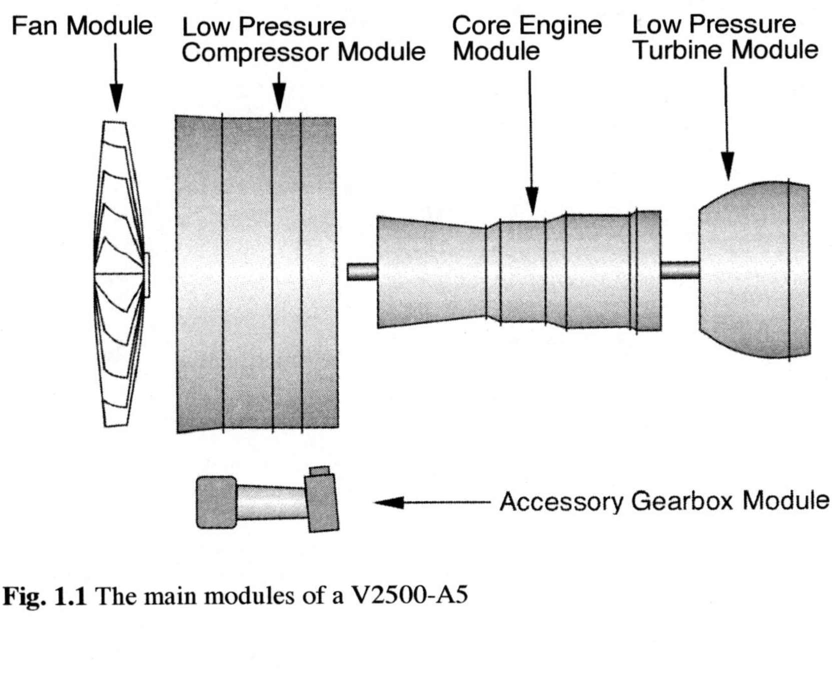

Fig. 1.1 The main modules of a V2500-A5

module can be removed from the engine without disassembling it into its piece parts. For the assembly process of the engine the modules are pre-assembled and then assembled to the complete engine. Each module has its specific maintenance schedules and as well as skill and tool requirements.

Besides reflecting maintenance aspects this modular concept is likewise useful to describe elementary functions of engine operation and incidences of the working cycle. In the following this compound of modules will be referred to as the basic engine.

The basic engine by itself is not operable and cannot serve all the functions the airframe depends on. Additionally to its main components the basic engine needs various systems to become an operable engine (Fig. 1.2).

These engine-related systems are usually called engine systems. To understand how an aircraft engine operates, it is not enough just to know how the basic engine operates. It is also essential to know how the engine systems work. This book has been written to introduce the reader to the basic design and the operation principles of the mainly used system designs on large turbofan engines. It gives an overview of the most frequently used system designs. To become familiar with a specific engine and its systems, it is always necessary to study the description belonging to this engine type.

The engine also provides the energy for airframe related systems (or simply airframe systems) like the pneumatic system, the electrical system and the hydraulic system. For the connection to these systems some hardware is installed on an engine. Other topics in the field of engine systems are the components necessary to turn the engine into a power plant. These are primarily the air inlet, the thrust nozzle and the cowlings, which are mainly structural components and no classical systems. But they belong to the group of engine systems.

1.1.1 Grouping of Engine Systems

For an overview of the engine systems they can be grouped in accordance with their function and importance for the engine operation. The first group consists of the primary engine systems. They represent the classical engine systems. These are:

- An air cooling and sealing system
- A lubrication system
- A fuel distribution system
- A control system
- Engine controls

- An indication system
- An ignition system
- An exhaust and thrust reverser system
- A starting system
- Cowlings, air inlet and nozzle

There are additional systems not belonging to the engine systems but they are functionally connected to the engine. These airframe systems, which are not active during normal engine operation but are activated to provide protection when required, can be grouped as the secondary engine-systems. These systems are:

- The fire protection system
- The ice protection system

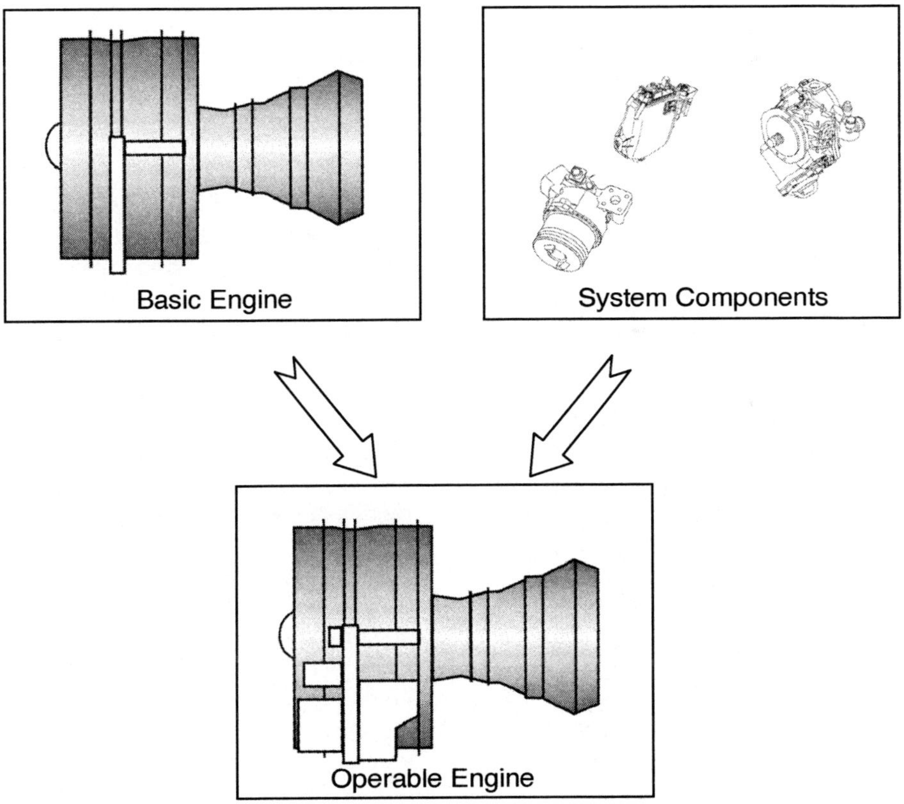

Fig. 1.2 By adding the engine systems to the basic engine it becomes the operable engine

The systems associated with the energy supply to the aircraft are grouped as the supply systems. They derive power, respectively air supply, from the engine but are also integral parts of their respective airframe system. These systems are:

- Electrical power generation system
- Pneumatic system
- Hydraulic power system

The reader may have the first contact with a technical description of an aircraft or engine type in the form of a training manual or a maintenance manual. In civil aviation it is international standard to group and name the systems in these publications according to ATA Specification 100 "Manufacturers´ Technical Data". This specification gives the chapter numbering standard for technical manuals used in civil aviation. In this standard each system of an aircraft has its own chapter with a standardized title and chapter number. The following Table 1.1 shows this in detail.

Table 1.1 ATA Chapter Numbers and Titles

ATA Chapter No.	Chapter Title (System)	Subjects to be found
70	Standard Practices - Engines	Standard Procedures to be observed while working on the engine
71	Power Plant	Air inlet, cowlings, engine mounts, power plant removal / installation
72	Engine	Description and inspections of the engine from the compressor to the turbine and the gearbox.
73	Engine Fuel and Control	Fuel distribution system, engine control system and fuel indication.
74	Ignition	Engine ignition System
75	Air	Engine air systems for cooling and compressor control.
76	Engine Controls	Engine controls and its components.
77	Engine Indication	Engine indications for power (EPR or Torque), RPM (N1, N2, N3), temperature (EGT, Nacelle Temp.) and vibration.
78	Engine Exhaust	Nozzle structures and thrust reverser system.
79	Engine Oil	Engine lubrication system and the related system indications.
80	Starting	Engine starting system.

Table 1.2 Secondary systems and supply systems

ATA Chapter No.	Chapter Title (System)	Subjects to be found
26	Ice and Rain Protection	Anti-ice systems for airframe and engine
30	Fire Protection	Fire detection and extinguishing for airframe and engine
24	Electrical Power	Elec. Power generation and network
29	Hydraulic Power	The complete hydraulic system
36	Pneumatic	Pneumatic system for the distribution of the pressurized air

This principle ensures an identical manual structure for every aircraft. The relevant chapters for the engine systems of turbofan engines have the number range from 70 to 80 of the system group Power Plant.

This group of chapters covers only the primary systems of the engine. The chapter numbers assigned to the secondary systems and the supply systems are listed in Table1.2.

If a description of engine systems were structured like a manual, this would not be the optimum approach for an understanding of the way they are linked. The following chapters are ordered in a sequence that the preceding chapters give as much basic knowledge as possible for the understanding of the following chapters.

The system descriptions in this book do not show the operating principles of every system component in detail, because some of this information is basic engineering knowledge (e.g. the operation of a gear type pump or a filter).

A system influences the operation of the engine main components due to cooling, air extraction or the fuel delivery for combustion. These influences are discussed only if the function of the described system is affected.

1.1.2 Demands on Engine Systems

Besides serving their allocated functions the engine systems must be designed to satisfy further objectives of the operator of the aircraft. These objectives are to provide safety and reliability, must be economically viable and competitive and meet certification requirements. These certification requirements include those for noise and pollution.

Providing safety and reliability requires introducing failure management into the design of a system i.e. avoiding failures and securing continuation of a flight even when a component has failed.

Avoiding operating errors can be accomplished by reducing the pilot's workload, through automation of systems, by providing limit protection and informing the pilot about system malfunctions and proposing appropriate responses. Failures of system components and of the total system can also be avoided by designing robust and fail-proof components.

If the reliability assessment of a component reveals weaknesses, other modes of operation must be provided in the event of a component failure. This is usually done by providing redundancy. Redundancy may also serve to increase dispatch reliability.

Efficient means of improving economy of operation are the reduction of weight, size and acquisition cost and optimizing operation with respect to fuel consumption and operational life.

A significant contribution to a safe and economic operation can be derived by proper maintenance. The design of a modern engine is, to a considerable extent, determined by its maintenance concept. E.g. the modular design of system components and easy accessibility will allow to replace components that have a higher failure probability in the limited time frames available when the aircraft is in line service.

A valuable contribution to an effective maintenance program is the continuous recording of data of the status and operation of all systems. The evaluation of this information allows to perform:

- System diagnosis
- Trend monitoring
- Failure and maintenance action prediction

As a consequence the systems of modern engines are provided with a multitude of sensors. The data sensed are not only used for the control of the engine or the system but also for the monitoring and analysis of the system condition. The data are fed into a central processing and data storage module. Some analysis is done by the on-board equipment, and further calculations are performed by ground-based computers.

Control of the systems is predominantly performed by embedded software which operates in real time and which interacts with the hardware functions. The software interprets input data from the sensors or data buses directly, processes the information and drives electrically commanded output devices.

This control system depends on a reliable data transfer and a reliable electrical power supply system. Software is also used to provide system monitoring, health diagnostic and fault monitoring information, which is processed for indication to the flight crew and stored for maintenance purposes.

Thus all systems of a modern engine are highly integrated and perform complex and interrelated functions within a federated system of data generation and transfer. Control and data processor functions are not performed in each system itself but in a central processor.

Thus when investigating a specific system it is important to notice that the specific functions of the system have not been the only design objectives. Safety, reliability, efficiency and maintenance functions are also important. A system meets these objectives best as a member of a combination of federated systems, which are generally controlled by embedded software and which exchange data with other system computers.

There are several textbooks providing the reader with the knowledge of the mechanical as well as aerodynamic and thermodynamic aspects of turbine engine designs. It is not the intention of this book to repeat what is already documented elsewhere. It is even expected that the reader has a general knowledge of the design and operation of a turbine engine that he can find in these textbooks. The design principles provided there will be relevant for many future generations of turbine engines. But as has been demonstrated, there has been an evolutionary change in the design of the last generation of modern turbine engines driven by reliability, economic and maintenance considerations. This evolution has been made possible by introducing additional new technologies that have become available by introducing data processing into engine design. These evolutionary changes become especially apparent when comparing the engine systems of older and the newest turbine engine designs.

In most relevant text books the systems of the jet engines are not covered in much detail. The influence of the new technologies discloses itself mainly in the design of the systems and systems architecture, which are not adequately represented either. The new technology systems have become a dominant factor in the operation and economic viability of the modern turbine engine.

Thus, as stated at the beginning, it is the intention of this book to give an introduction to the systems of modern civil turbofan engines. These demonstrate a significant change compared to respective older designs. The introduction to the systems will be given by investigating the systems of some of the current generation of turbofan engines powering the present generation of large commercial transport aircraft. As these systems proved to be very successful, their design principles have also been adopted in the newest generation of smaller turbofan engines used in regional and business aircraft.

This investigation concentrates on the primary systems that support the operation, control and supply of the basic engine. The supply systems (pneumatic, electrical power generation and hydraulic power) will only be covered in so far as they interact with the engine.

Generally speaking, an insight into the design and operation of the systems and the interoperability with other systems is given. No methods to design a system or to analyze the operation of a specific system are presented.

1.2 Digital Aircraft Systems

1.2.1 General

The reader may ask why an introduction to digital aircraft systems makes sense in a book about engine systems. It makes sense because transport aircraft have undergone significant changes during the last decades with the adoption of the digital technology available during their design phases. Digital technology has been applied to individual systems throughout the aircraft, including the engine, allowing them to become more reliable, lighter and easier to maintain. Thus a general description of the digital technology of aircraft systems seems to be justified.

These systems also provide more functionality than their predecessors. Onboard computers now control many of those functions that were controlled by pilots and flight engineers on aircraft of older designs. A flight engineer is no longer required and the pilot's systems related function has changed from a system controller to a supervisor.

The computers controlling the aircraft systems and providing communication between these systems generally share the same technology. Thus the expression *avionics*, originally used only for systems like communication and navigation, is also used for the digital components of other systems.

A computer embedded in a hardware component and interacting with the system may e.g. operate as a closed loop controller. It continuously samples sensor outputs, calculates appropriate control responses and sends those responses to actuators. In safety critical applications multiple installations of sensors and computers are available and the control loop becomes more complicated by the addition of a redundancy management and synchronization and voting mechanics. Thus basic elements for a system controller are sensors, a processor, actuators and communication systems for data exchange with remote sensors and other aircraft systems.

1.2.1.1 Sensors

Sensor output quality is a significant factor for control functions, cockpit indications and data collection for monitoring and maintenance support. The introduction of new technologies as e.g. microelectromechanical systems (MEMS) in combination with modern microprocessors has greatly improved sensor output quality.

Sensors that are used in digital control and data acquisition systems must provide an electric signal in digital format. Generally sensing elements, e.g. electro-mechanical systems (MEMS), produce an analog output signal. Depending on the output quality of the output signal it is directly fed into an analog/digital converter or it is amplified and filtered before entering the analog/digital converter.

The converter output is fed into a microcomputer in which several algorithms are implemented that may serve the following functions:

- Compensation for fabrication imperfections
- Compensation for temperature and non-linearities
- Digital filtering

Built-in automatic test functions provide both continuity and power checks to examine the integrity of the device.

The microcomputer can be configured to transmit its output as an analog signal or in a standard serial communication protocol such as RS-232, RS-422, RS-485 or ARINC 429. Bidirectional communication allows the user to calibrate the sensor without removal. Signals from a number of such sensors can be multiplexed and transmitted on a single pair of wires.

Not all sensors use all the features presented here but most of these features are available on pressure transducers used for such safety critical applications as FADEC systems and air data computers. The above-mentioned properties e.g. are found on the "smart" pressure transducer Honeywell LG-1237 of which up to nine are installed in the CFM56-5C engine's FADEC computer.

The sensor/microprocessor unit will have to communicate and thus interface with the embedded system. Interface standards will be described in more detail in Appendix A.

1.2.1.2 Processor

The capabilities a processor must provide are determined by the applications it has to serve. There is a clear tendency to replace, wherever possible, mechanical functions by software functions. E.g. the FADEC computer installed on modern engines weighs in the order of 15 kg. It does not only replace a hydromechanical fuel control of older engines, which weighs in excess of 50 kg on large engines, but also provides many more functions and a higher reliability. Embedded computers may incorporate hardware well known from PCs, but they differ in the operating system they use.

American and European regulators have published documents RTCA/DO-178B respectively EUROCAE ED-12B that define guidelines for the development of safety-critical software for the use in airborne systems. These documents comprise five levels of safety-criticality with level A requiring the most demanding certification process.

Because an airborne system will only be certificated if its software meets the standard, special operating systems have been developed. They simplify the certification process.

Besides supporting safety-critical applications the operating systems must be generally able to support time-critical applications. There are two time-critical applications:

- Time Sharing
  Under all operating conditions a high priority computation is completed within a predetermined time frame. Closed loop controllers, e.g. the fuel metering controller of an engine, require that sampling sensor output and computing control responses are continuously fully completed within a defined time frame.
- Event Driven
  If a message has been received or a computation has identified a certain condition, a response is triggered within a predetermined time frame. Such a system may trigger an alert message, may activate envelope protection or the ignition system.

Operating systems providing time-critical services are called real time operating systems (RTOS). They do not actually perform real-time computations but guarantee that predetermined deadlines for the execution of computations are met. These are multi-tasking systems. They incorporate task schedulers, which make sure that high priority tasks are continuously fully completed within the set time frame.

An embedded system serving as an element for a control loop will interact with its hardware by transmitting signals to the actuators like fuel metering devices, hydraulic actuators, solenoid valves etc. Generally these require analog signals for their operation. Thus digital/analog converters are installed to provide the appropriate command signals. Further details about the diverse types and functions of actuators will be given in connection with the description of the systems to which they belong.

Many of the aircraft systems must communicate with each other. The engine control system, e.g., may require thrust demand signals from the autothrust system. This data transfer is mainly performed with the help of data buses.

1.2.2 Federated Avionics Architecture

Data communication between safety critical systems implies the danger of failure propagation from one system to another. One strategy satisfying certification standards is arranging safety critical applications in a federated architecture. This means that each function (e.g. autothrust) has its own fault tolerant computer with only limited interconnections to systems of other functions. Interactions are limited to the transfer of data. Functions are available which detect or allow to tolerate faulty data or isolate a faulty data source. This is an effective means of preventing fault propagation, because the systems providing different functions do not share any resources. The failure of one function has little effect on the continued operation of others. The partitioning of functional capabilities into separate computers allows to provide appropriate levels of redundancy offering a maximum of safety and dispatch reliability.

A bus system appropriate to federated avionics architecture is the ARINC 429. Transport aircraft using the ARINC 429 data bus in a federated avionics architecture are the Airbus A320, A340 and the Boeing 757/767.

1.2.3 Integrated Modular Avionics

The obvious disadvantage of the federated avionics architecture is the amount of hardware required, because each function requires its separate computer. Some of these have to be replicated to meet redundancy requirements. Additional disadvantages are their weight and the requirements for space, power, cooling, installation and maintenance.

Taking advantage of recently emerged computer technologies, integrated modular avionics (IMA) architectures have started to replace federated avionics architectures.

The expression "modular" indicates that standardized architecture and components are used. The expression "integrated" implies that some of the components are shared by different functions, even functions of different criticality.

In IMA architecture one fault tolerant computer system provides a common computing platform for multiple functions. But integration of multiple functions on a single processor may lead to fault propagation and intransparent behavior. Because individual functions cannot provide protection against corruption to the computer on which they depend, the IMA architecture must provide means of avoiding fault propagation from one function to another. In an IMA architecture partitioning provides protection against fault propagation. Partitioning is performed by specific operating systems. They provide time, resource and health management. Each partition has a fixed and predetermined allocation of time and memory. This allows secure separation of functions and information within a system so that one function cannot, intentionally or unintentionally, interfere with the operation of another function. Its information cannot be accessed by an unauthorized application. This allows applications with different levels of criticality to use the same hardware. Partitioning provides the opportunity to introduce modular architectures in the hardware and software and thus to use concepts already widely applied in PC and commercial off-the-shelf (COTS) technology.

Interchangeable hardware avionic modules facilitate technology upgrades within a system when their components become technologically obsolete or are no longer in production. Hardware computer components are often replaced within only a few years whereas aircraft systems may be in service for several decades.

This concept requires that the software can be moved to a new computer with no or minimal rework. Modular software will allow to divide complex software into separate building blocks to fulfill this requirement. Further advantages are that improved versions of independent building blocks can be developed and testing is simplified when individual building blocks have been tested beforehand.

Traditionally software has been certified only as part of an avionic hardware component and not separately. But now FAA AC-20-148 provides guidelines for the reuse of even safety-critical software, which has already been tested and certified in another project. This will allow to develop libraries of tested and certified reusable software components (RSC) for multiple use.

The first application of the integrated modular avionics architecture was introduced in 1995 on the Boeing 777. Its Aircraft Information Management System (AIMS) is a "cabinet" that houses standard line replaceable modules (LRMs) for data processing, input/output processing and power supply. These cards are connected to a back plane, which connects the modules by means of a data bus. The cabinet consolidates the data processing for 10 aircraft systems including flight management, thrust management and central maintenance. For bi-directional data transfer with other LRUs an ARINC 629 data bus network is used. More information concerning data busses is provided in Appendix A.

The Boeing 787 and the Airbus A380 adapt IMA architectures but demonstrate different approaches. The B787 uses two redundant central computing systems called common core system (CCS), which replace about 100 traditional LRUs. In contrast the IMA architecture of the A380 relies on eight processing modules. The A380 and the B787 use AFDX networks for data transfer. More details about this can also be found in Appendix A.

The IMA hardware architecture and operating system do not extend to all aircraft systems. A few flight critical systems employ separate hardware components and operating systems.

Besides, the engine control system is not integrated into the airframe system architecture, although it communicates via an AFDX data bus with the airframe systems. The main reason is that an engine has to be developed and certified separately from the airframe and may be mounted to different airframe types. Furthermore, the engines are generally far away from the avionics bay, which also favors a local controller. Nevertheless, principles of IMA architectures like running functions of different criticality (e.g. engine control Level A and maintenance data collection Level D) on the same processor have also been applied to engine control systems.

1.3 Aircraft-Engine Interface

1.3.1 Structural Interface

An engine installed to an engine pylon (or strut) has a structural interface to the airframe with its attachments and the cowlings. The load carrying structure of the engine is attached to the airframe by the use of adapters called engine mounts. These mounts are installed on the engine cases and then attached to the adapters on the engine pylon.

The cowlings that cover the engine have a structural interface to the engine pylon. They are attached with hinges to the pylon, which carries the weight of the cowling. The outer surface of the cowlings has a smooth transition to the surface of the pylon to ensure minimum airflow disturbances.

1.3.2 System Interfaces

Several aircraft systems receive their energy from the engine. Thus components related to these systems must be installed on the engine to connect these systems to the engine.

For the transfer of data between the engine control system and the aircraft systems some hardware is necessary and must be installed on the engine. Also the appropriate software must be installed in the engine control system and in the assigned computers within the airframe.

More details about the structural and system interfaces can be found in Chapter 10. The interfaces for data transfer are described in Chapter 5.

1.4 Thermal Management

1.4.1 Thermal Management for Engine Parts

During engine operation the heat of the main gas flow warms up the engine parts. The parts that are in contact with the hot combustion gas can become so hot that their material temperature could exceed the limit established for a long life of a part. To ensure a material temperature below this limit the parts are cooled. This means some of the heat is removed from the parts by a cooler medium, usually air from the compressor.

To guide this cooling airflow to the part that must be cooled, an internal air system is established within the engine. The cooling effect of this system matches with the cooling demand of the parts to be cooled and ensures material temperatures below the limits.

Another problem arising with heat is the thermal growth of engine parts. The engine parts that are matched in the cold condition must also match at operating temperature. Therefore the thermal growth of matched parts must be managed. This can be done by cooling the relevant part or by an appropriate design in consideration of the right material combinations. Thus the thermal growth of a rotor and a case can be matched sufficiently.

An example of the matching of thermal growth by cooling is the case cooling of the turbines. More information about all these cooling systems can be found in Chapter 2.

1.4.2 Thermal Management for Fuel and Oil

For fluids a thermal management is necessary, too. The engine oil is the fluid in the engine that can become too hot. It absorbs heat from the hot engine parts it has contact with. To prevent excessive oil temperatures, the oil must be cooled. The fuel is the fluid that can become too cold. It usually cools down in the wing during the flight and needs to be heated before it enters the engine fuel system.

Fuel is able to absorb a large amount of heat. Thus fuel is usually used to cool the oil of the engine. The result of this is, that the requirement of heating the fuel is fulfilled and the necessary oil cooler has small dimensions.

The simplest cooling system comprises one fuel-cooled oil cooler for the cooling of the engine oil. The size of this cooler matches with the cooling demand of the oil and the oil temperature is kept within the limits during the whole operating range of the engine.

If the fuel also cools the oil of the engine-driven aircraft generator, the balance of oil cooling and fuel heating can be disturbed under certain operating conditions by the additional heat of the generator oil. To manage the temperatures of engine oil and fuel in such a system a more complex thermal management system is necessary. These systems are operating with additional fuel valves to redirect the flow of the cooling fuel, with an additional air-cooled oil cooler or with both features. If an air-cooled oil cooler is used, the system minimizes the use of it and therefore the use of cooling air from the gas path. More detailed information on these systems can be found in Chapter 4.

1.5 Sample Engines

The sample engines mentioned in this book are engines operated in large numbers. The reader has a good chance to get in touch with some of these engines in the real world of aviation. It is not the intention of this book to give you a type training on one or two engines used as samples.

To show you a wide range of used technical solutions, the examples are taken from the engine type that best pictures the discussed subject. The engines selected are installed in aircraft with more than 100 seats. For engines used in smaller jets and business aircraft the same basic principles apply.

Table 1.3 shows a listing of the sample engines with its basic characteristics. Additionally the aircraft applications and the manufacturers are listed. Except for the marked engines all these engines are controlled by a full authority digital engine control (FADEC) system. If details of the PW4000 are shown in one of the following chapters, the 94-inch version is meant.

Table 1.3 List of Sample Engines (* hydromechanical fuel control)

Engine	Manufacturer	T.O. Thrust Range	Fan Diameter	Application
CF6-50*	GE Aviation	224 - 240 kN (51000 - 54000 lbs)	2.2 m (86.4 in)	B747-200 DC10
CF6-80C2*	GE Aviation	233 - 282 kN (52400 - 63500 lbs)	2.36 m (93 in)	A300-600 A310
CF6-80C2B1F	GE Aviation	233 - 282 kN (52400 - 63500 lbs)	2.36 m (93 in)	B747-400
GE90	GE Aviation	329 - 512 kN (74000 - 115000 lbs)	3.25 m (128 in)	B777
GEnx	GE Aviation	236 - 334 kN (53000 - 75000 lbs)	2.82m (111in)	B787 B747-8
CFM56-3*	CFM	82 - 105 kN (18500 - 23500 lbs)	1.52 m (60 in)	B737-300/400/500
CFM56-5A	CFM	98 - 118 kN (22000 - 26500 lbs)	1.73 m (68 in)	A319, A320
CFM56-5B	CFM	98 - 147 kN (22000 - 33000 lbs)	1.73 m (68 in)	A318, A319, A320, A321
CFM56-5C	CFM	139 - 151 kN (31200 - 34000 lbs)	1.83 m (72 in)	A340-200/300
CFM56-7B	CFM	87 - 121 kN (19500 - 27300 lbs)	1.55 m (61 in)	B737-600/800/900
V2500-A5	IAE	98 - 147 kN (22000 - 33000 lbs)	1.61 m (63.5 m)	A319, A320, A321
JT9D-7R4*	Pratt & Whitney	214 - 249 kN (48000 - 56000 lbs)	2.37 m (93.4 in)	B747-300, B767 A310, A300
PW4000	Pratt & Whitney	231 - 276 kN (52000 - 62000 lbs)	2.39 m (94 in)	B747-400, B767, A310
RB211-524*	Rolls-Royce	222 - 270 kN (50000 - 60600 lbs)	84.8 in	B747-200 B747-300
Trent 500	Rolls-Royce	236 - 267 kN (53000 - 60000 lbs)	2.47 m (97.4 in)	A340-500 / 600
Trent 700	Rolls-Royce	300 - 334 kN (67500 - 75150 lbs)	2.47 m (97.4 in)	A330

1.6 Definitions and Terms

1.6.1 Gas Path Stations

Reference planes within the gas path are called gas path stations. In a gas turbine engine the stations correspond to the beginning and the end of thermodynamic processes in the engine. For the designation of these stations in the gas path of an engine a designation system is used.

To make the designations of the stations as easy as possible, the stations are designated with numbers of a standardized numbering system. This designation principle makes technical reports, documents, and conversations much more concise and easy to understand because all engines using this standard have the same station numbers at the same points in the working cycle. This standard is provided by the SAE Standard AS755 Rev. D. Figure 1.3 illustrates the principle of the station numbering according to this standard.

Basically a gas path station can be established in front of and behind every airfoil row and not only at the characteristic points of the working cycle. For practical use they are established at locations that are important for maintenance. These locations are inlets or outlets of main components like compressor, combustion chamber or turbine representing the beginning or end of working cycle sub-processes.

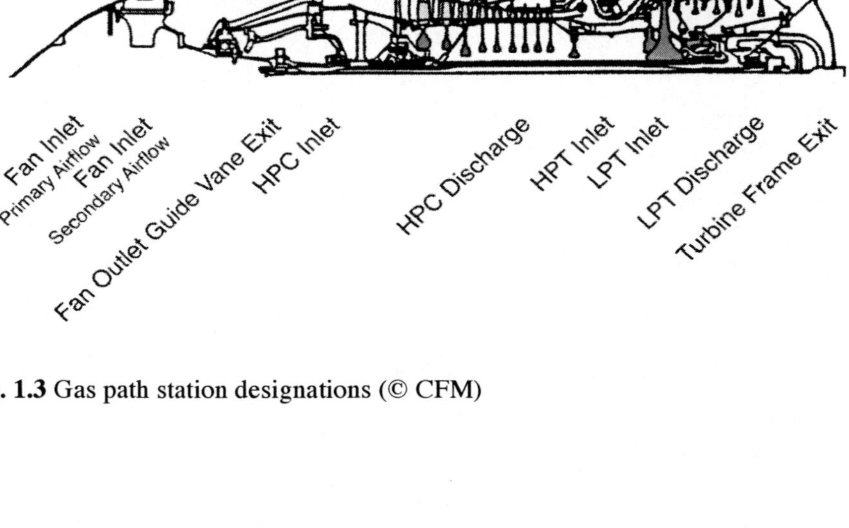

Fig. 1.3 Gas path station designations (© CFM)

The temperatures or pressures sensed at a station are designated using the symbol for the sensed value and the station number. Pressures can be sensed as total pressures and static pressures depending on the design of the pressure probe. A second letter in the pressure designation indicates this difference. This letter can be written as a subscript index letter or in line with the P. For example the static pressure at station 12 is expressed as Ps12. A total pressure at this station would be expressed as Pt12. For simplification the total pressure designations are often written without the index letter. Thus Pt12 and P12 mean the same.

Temperatures can only be sensed as total temperatures. Thus the index letter "t" is normally not used. Temperature designations like T12 or Tt12 mean the same.

Ambient pressures or temperatures have the station number 0. The pressure P0 is a typical example. It is used to calculate the actual altitude.

The sensors installed at certain locations are named according to the stations where they are installed and the value they sense. Typical designations are T2 Sensor or T2/P2 Sensor for example. The latter is a combined sensor with a temperature and a pressure probe.

1.6.2 Pressure Ratios

For the evaluation of the compressor and turbine efficiency several pressure ratios are defined. By using the station designations explained before they are expressed in simple terms. For the compressors these are:

- Fan Pressure Ratio = P13/P12
- HPC Pressure Ratio = P30/P25
- LPC Pressure Ratio = P25/P20
- Compressor Pressure Ratio = P30/P20
- Overall Pressure Ratio = P30/P0

For the evaluation of the pressure drop across the turbines the following pressure ratios are used:

HPT Pressure Ratio = P30/P45
LPT Pressure Ratio = P45/P50

For the determination of the thrust produced by the engine *the engine pressure ratio* (EPR) is calculated and indicated on the flight deck. The EPR is defined as:

EPR= P50/P20

The EPR is an equivalent value to the engine thrust and the engine thrust setting can be expressed in EPR values. For the calculation of the EPR the total pressure at the fan inlet and at the nozzle inlet are required. On engines with control systems using the EPR as the control parameter for thrust control the pressure probes for P20 and P50 are installed in the corresponding cases. The assigned computer calculates the EPR value.

1.6.3 Shaft Speed Designations

The rotational speeds of the engine shafts or rotors are designated with the letter N and the number that represents the position in the sequence the compressors are reached by the airflow. E.g. the low pressure compressor is the first compressor reached by the airflow and the low pressure rotor speed is therefore called N1. The high pressure rotor speed is designated N2 respectively. In a three-spool engine N2 is used for the intermediate pressure rotor speed and N3 for the high pressure rotor speed. Note that the letter N is usually written as a capital letter.

1.6.4 Corrected Parameters

The parameters measured in a turbine engine at a certain thrust setting are dependent on the actual atmospheric conditions. Thus the parameters have different values when they are measured under different atmospheric conditions but at the same thrust setting.

The condition of the atmosphere is determined by the values of its temperature, pressure and humidity. When measurements are made, which are dependent on these values on different days and/or at different locations, these sensed values differ between these measurements. To make the values of all measurements comparable, they are converted to the same atmospheric conditions. These are the conditions of the International Standard Atmosphere (ISA). The International Standard Atmosphere was created to make data comparable that are dependent on the atmospheric conditions. The most important conditions of the ISA at sea level are:

Pressure	1013.25 hPa
Temperature	15°C (288 K)
Density	1.225 kg/m^3
Humidity	0%

The ISA conditions are also known as the standard day conditions. Parameters measured under conditions different from ISA can be converted to ISA conditions. This means that the result gives the value the parameter would have had if measured under ISA conditions. Such a parameter converted to ISA conditions is called a corrected parameter or a standard day parameter. It is expressed with the index k or corr. For a pressure value this would show P_k or P_{corr}.

For the efficient storage of data in the databases of the fuel control system the characteristic diagrams are stored for the ISA conditions only. The data are converted to the actual atmospheric conditions by the respective calculations established in the software.

1.6.5 Location Designations on Cases

For the description of a position on an engine or an aircraft basically a view from the aft into flight direction is assumed. Directions like "left of centerline" or "right of centerline" are based on this viewing direction. If another viewing direction is assumed, it must be stated to prevent misunderstandings

For the identification of a location on a circular engine case for maintenance purposes the clock time positions are used. Only on manufacturing and repair drawings precise angle values are used. Using the clock time positions assumes the 1200 o'clock position at the top dead center of the case and a view from the rear into flight direction. Figure 1.4 shows an

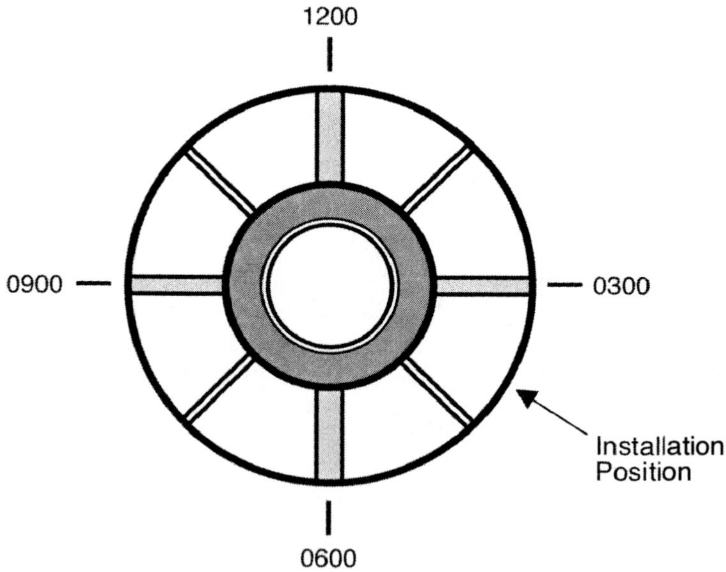

Fig. 1.4 Clock positions on a circular engine case with an example of an installation position at approximately 0400 o'clock

example of a circular case with an installation position of a component marked at the 0400 o'clock position.

If a viewing direction against the flight direction is assumed, it is stated in the text or on the respective drawing with the words "Forward Looking Aft" (FLA). In some cases it is useful to state the viewing direction from the rear to prevent misinterpretation of the drawing. This is done with the words "Aft Looking Forward" (ALF).

2 Engine Air Systems

All systems dealing with cooling air, sealing air or compressor control, belong to the engine air systems. The bleed air system for the supply of the pneumatic system of the aircraft does not belong to the engine air systems. It is a part of this aircraft system. To get an overview of the various engine air systems, it is useful to divide them into the following two main groups: The internal air systems and the external air systems.

2.1 Internal Air Systems

The numerous cavities inside the rotors and between the rotor components and the cases are used to guide the airflows of the internal air systems. The air is bled from the main airflow at selected compressor stages into these cavities for different purposes. Because the temperatures and pressures needed are unique to each engine type, the internal air systems are always different from one engine type to another. The different designs of the internal components also lead to different internal air systems. To keep the description as simple as possible, here the examples of two-spool engines are described.

2.1.1 Component Cooling and Sealing

A large portion of the internal airflow is used for the cooling of the engine's rotor components. Following the airflow from the front of the engine, the need for cooling the rotor components begins in the area of the high pressure compressor (if we look into a two-spool engine). The air enters the inner cavity of the high pressure rotor at the front end of the HPC or at one of the higher stages of the HPC. Inside the rotor drum the air flows towards the rear end of the spool and also cools the HPT rotor disks in its bore area. After this cooling flow has left the high pressure spool it is used to cool the disks of the low pressure turbine before it leaves the rotor cavities towards the main gas flow. Fig. 2.1 shows such an airflow.

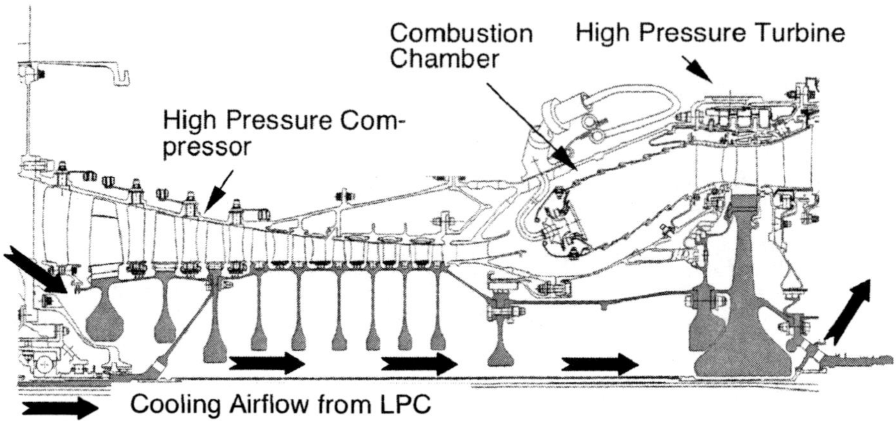

Fig. 2.1 Internal cooling airflow in a CFM56-7B (© CFM)

To get the right cooling effect for a certain engine component, air with the proper temperature and pressure is needed. Therefore the designer must carefully select the right compressor stage as the air source for a cooling purpose.

At the last compressor stage air from the main airflow can enter the cavity between the high pressure rotor and the combustion case. To prevent a leakage from the main airflow, a seal is often installed between the high pressure rotor and the combustion case. It is called CDP seal or thrust balance seal. Fig. 2.2 shows a CDP seal of the labyrinth type. It can also be designed as a brush seal. Both types of seals have in common that only a small amount of air can pass across these seals. Thus a small airflow exists across the CDP seal into the inner cavity of the combustion case and the air pressure here is lower than the pressure at the compressor discharge.

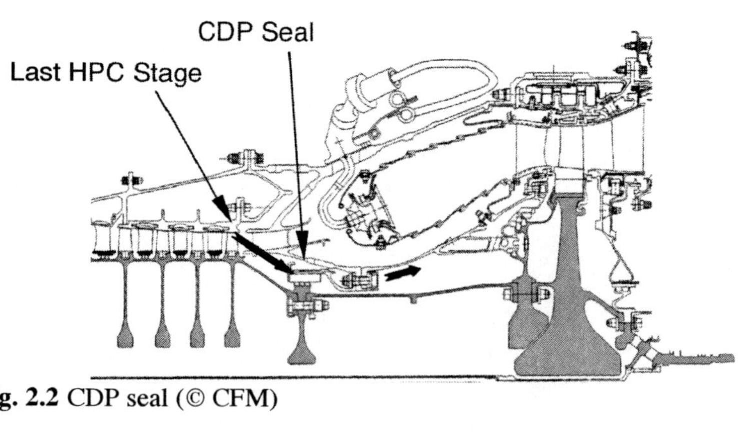

Fig. 2.2 CDP seal (© CFM)

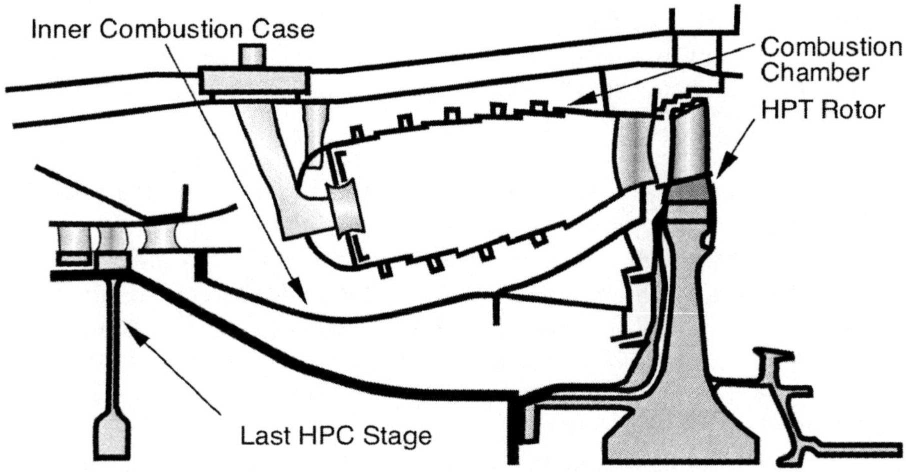

Fig. 2.3 HPC exit without CDP seal of the Trent 500 (simplified)

This air is then used for sealing and cooling purposes at the high pressure turbine. If a bearing compartment is located within the center of the combustion case, this air is also used to seal this bearing compartment. On some engine designs, as shown in Fig. 2.3, no CDP seal is installed. Here the air pressure from the exit of the last HPC rotor is present at the front face of the HPT disk.

A bearing compartment in the inner cavity of the combustion case needs cooling to keep the temperature of the compartment walls low for the prevention of oil coking. Figure 2.4 shows the supply of cooling air to this compartment. The air is delivered through an external system via an air cooler. After this air has passed the cooling air passages around the bearing compartment it exits into the internal airflow of the combustion case.

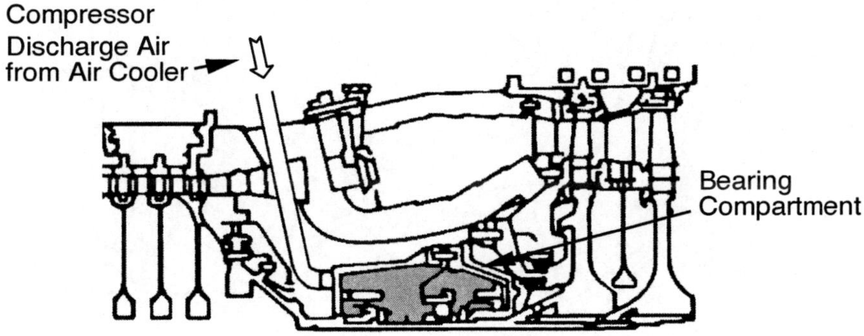

Fig. 2.4 Center bearing compartment of a V2500

The next important area for cooling is the high pressure turbine. The high pressure turbines used have no more than two stages. Here normally the blades and the nozzle guide vanes of both stages are air-cooled. Fig 2.5 shows the cooling of a PW4000 HPT.

For the first stage the air from the combustion chamber secondary flow is used. The cooling air part of this flow enters the nozzle guide vanes directly behind the combustion chamber. To cool the rotor blades the cooling air is directed through a ring of air nozzles in the static combustion case towards the rotating high pressure turbine. To guide the cooling air within the rotor to the blades of the first stage, most turbine designs use the cavity between the rotating air seal and the HPT disk on the front side of this disk. So the cooling air can flow through this cavity radially outwards to the blade roots where it enters the blades.

For the cooling of the second stage blades cooler air from the last compressor stage is used. Behind the CDP seal this air flows through the internal cavity of the combustion case. From here this cooling air enters the cavity between the turbine disks through holes in the first stage turbine disk and flows through the cooling passages of the blades.

The cooling air for the nozzle guide vanes of the second turbine stage is supplied from a more forward compressor stage than the cooling air for the first turbine stage. The air is delivered to the stator through externally mounted tubes. On some engines this cooling air supply can be switched off partially during cruise to save fuel. In this flight phase the gas temperatures are lower than during high power operation.

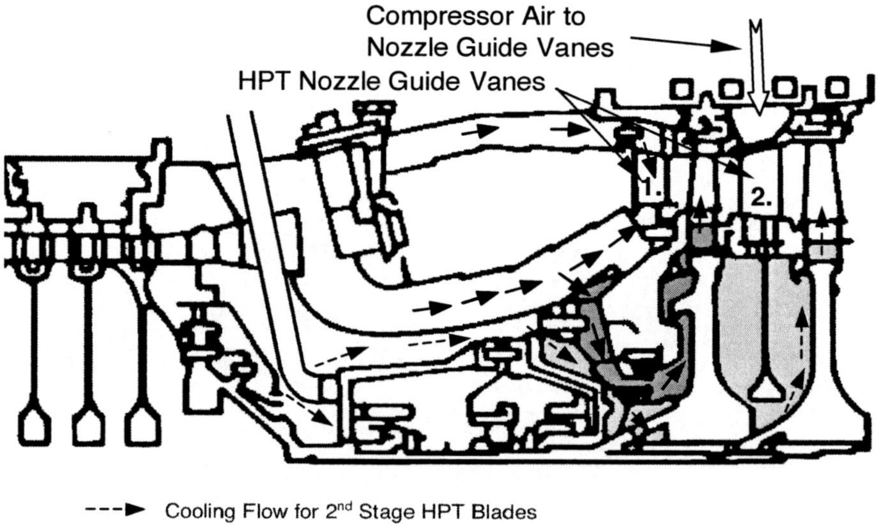

Fig. 2.5 HPT cooling airflow in a V2500

At the front and rear end of a turbine rotor it is necessary to seal the gap between the rotor and the static structure. Here labyrinth type seals are installed. The air for these seals originates from the inner cavity of the combustion case and from the high pressure rotor.

2.1.2 Pressure Balancing

The compressor and the turbine induce high axial forces on a rotor system of a turbine engine. The compressor induces a forward directed force and the turbine induces a rearward directed force. The resultant axial force is a forward directed load for the location bearing of the rotor system. To make the use of smaller location bearings on all shafts for the desired lifetime possible, the designers reduce this resultant load by the use of air pressure differences acting on the rotor. Figure 2.6 shows this principle for the N1 rotor as an example.

By applying a pressure difference across a rotor disk web or a rotor support, a rearward directed force is generated. Mainly turbine disks and turbine shaft components are used for this function. The remaining axial load for the location bearing is lower than without this pressure balancing, but it is not zero. This axial load must be kept high enough for the proper operation of the location ball bearing. In such a bearing the balls are relatively large because the axial loads are higher than the radial loads. Therefore the minimum axial load must be able to keep the balls rolling. Otherwise the balls can skid on the bearing races, which leads to bearing damage.

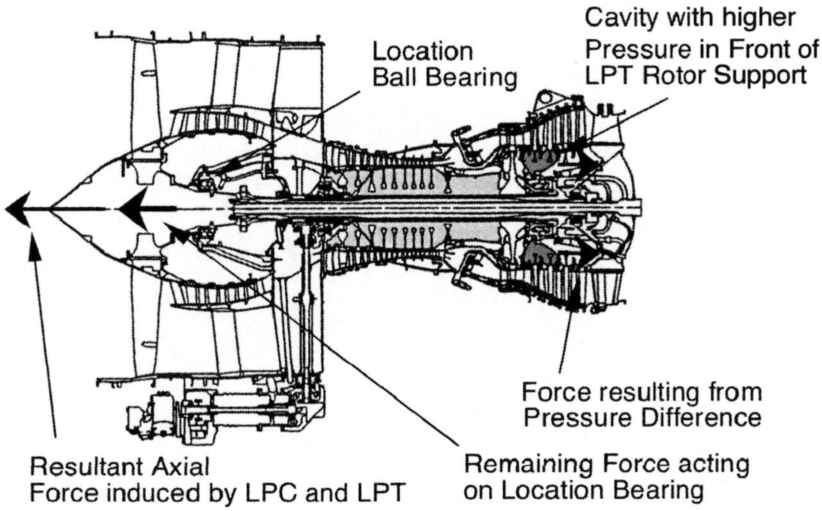

Fig. 2.6 Principle of pressure balancing (© CFM)

2.1.3 Bearing Compartment Pressurization

The term bearing compartment pressurization is known for the supply of sealing air to the bearing compartments. For the proper operation of the shaft seals between the bearing compartment wall and the shaft, sealing air must flow across the seal into the bearing compartments regardless of the type of the seals (labyrinth type or carbon seal). This sealing air leaves the main airflow usually between the two compressors to seal the front and rear bearing compartment. If a center bearing compartment is installed, the air behind the CDP seal is used to seal this bearing compartment.

To maintain this sealing airflow into the bearing compartments a venting of the bearing compartments is necessary. Therefore a vent system is installed as the third subsystem of the lubrication system. This vent system can be built from vent tubes for each bearing compartment routed through the cases. Or the low pressure shaft is used to house the vent tube (GEAE & CFM). Figure 2.7 shows this design variant. The vent air leaves the engine via the so-called center vent tube through the rear end of the low pressure shaft. So no external vent tubes are necessary.

All the other features of the vent system are described in Chapter 3, Lubrication System.

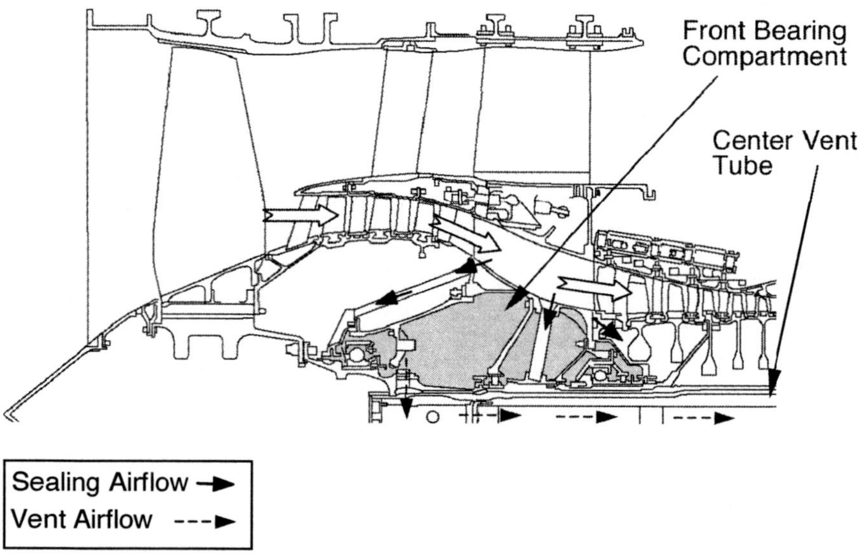

Fig. 2.7 Bearing compartment pressurization (© CFM)

2.2 External Air Systems

The air systems installed on the engine and necessary for the engine operation belong to the external air systems. In this group we find the cooling and ventilation system, the active clearance control systems and the compressor control systems.

2.2.1 Cooling and Ventilation Systems

This group of the cooling systems contains all the air systems for the cooling of engine components and accessories. For the component cooling simple tubes and hoses deliver air from the fan airflow to the cooled components. Figure 2.9 shows these tubes on the left side of a PW4000.

Of the ignition system it is usually the high tension lead that is air-cooled. On installations at the core engine the ignition exciter is air-cooled too. Figure 2.8 shows this configuration. The cooling air flows through a gap between the ignition exciter housing and a cooling shroud. The air leaves the cooling shroud through the cooling passage of the high tension lead. This cooling passage is located between the high tension insulation and the outer steel braid of the high tension lead.

The ventilation systems are used to deliver the necessary ventilation air into the cavities between the cowlings and the engine case. This ventilation is necessary to cool the components installed on the engine cases and to prevent the accumulation of flammable vapors in case of fluid leaks. The cowlings installed around the engine create two or more cavities that need ventilation.

Fig. 2.8 Cooling of the ignition system components (© LTT)

One is located between the fan cowling and the fan case; the other is located between the core cowling and the core engine cases including the LPT cases. The fan cowling cavity is ventilated with ambient air. This air enters the cowling cavity from the slipstream through air inlet ports and exits this area through air outlets 180 degrees apart from the inlet ports. The example in Fig. 2.9 shows this principle for the fan cowl ventilation.

The cavity inside the core cowling is ventilated with fan air from the fan air duct (secondary airflow). The air enters the core cowling cavity through scoops or simple holes in the core cowling and leaves this area either through a gap between the core cowl and the primary nozzle or through the latch area of the cowling in the 6 o'clock position. The amount of cooling air flowing through this cowling cavity depends mainly on the pressure difference between the fan duct and the ambient air.

To intensify the cooling during flight at low altitudes on some engines additional cooling air valves are opening to increase the fan airflow into the cowling. The airflow through such a system is shown in Fig. 2.9 as general nacelle cooling.

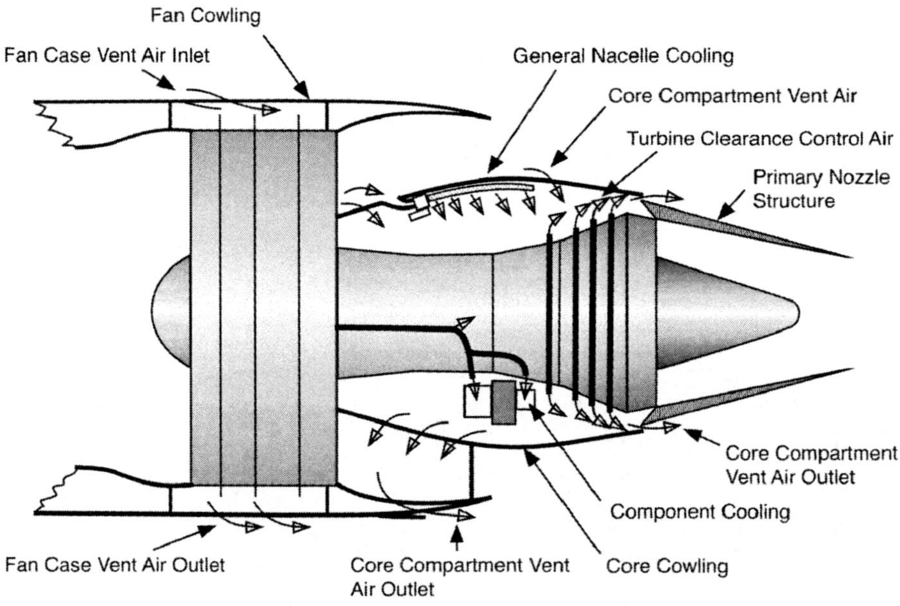

Fig. 2.9 Cowl ventilation and component cooling

2.2.2 Active Clearance Control Systems

The turbine engines operate with different gas temperatures that change with the power setting and the outside air temperature (OAT). The energy for the heating of the turbine rotor and the turbine case comes from the gas flow with its variable temperatures. Due to the different thermal expansion characteristics of the turbine rotor and the turbine case the clearance between the turbine blade tips and the turbine case (*tip clearance*) changes during the engine operation. A larger tip clearance decreases the efficiency of the turbine with the result of a higher thrust specific fuel consumption (TSFC) and higher gas temperatures because the lack in efficiency is compensated for by the fuel control system with more fuel. A too small tip clearance can lead to rubbing of the turbine blades on the case.

To prevent these undesirable effects a control of the thermal expansion of the turbine case is necessary. This is ensured by a cooling system that has a changeable cooling effect, the *active clearance control system* (ACC system). The cooling medium is air from the HPC or the fan duct that is impinged on the surface of the turbine case. Figure 2.10 shows the arrangement of the necessary cooling tube assembly. A control logic controls the system for the optimum tip clearance for the momentary operating condition of the engine. An additional effect is the reduction in the case temperature during engine operation. This extends the life of the case. To keep this cooling up during a system failure, some engines use LPT valves that allow a minimum airflow in the closed position of the valve for fail-safe operation. This prevents the thin walled LPT cases from operating with excessive material temperatures while the system is not controlled.

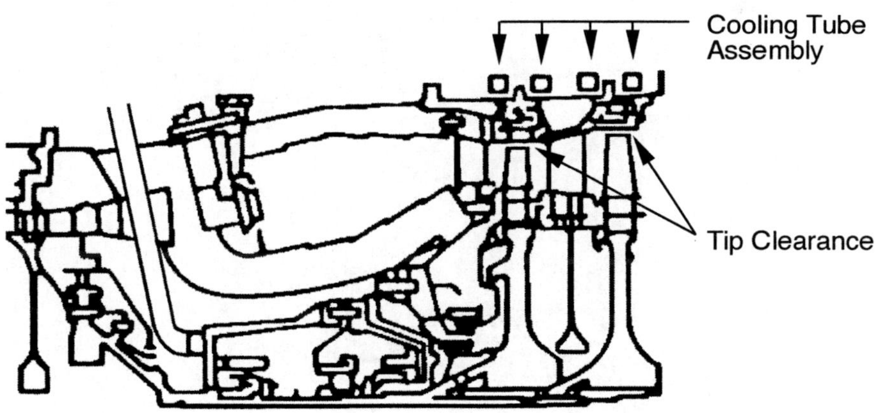

Fig. 2.10 Principle of turbine clearance control

The efficiency of the high pressure compressor has a great influence on the TSFC of an engine. So it makes sense to minimize its tip clearance for maximum performance. In most engines the HPC rotor has a permanent uncontrolled internal cooling flow. The expansion of the case matches up with that of the rotor so that the tip clearances become smaller at high rotor speeds and pressure ratios. The tip clearances must be designed with a margin to prevent rubbing because the cooling of the rotor is uncontrolled. The efficiency of the compressor can be increased if this margin is used for the reduction of the tip clearance during the flight phases of climb and cruise. This can be done by a system that heats the HPC rotor up during these flight phases with warm compressor air or with a system that shuts off the internal cooling airflow. The CFM56-5C is an engine with a clearance control system for the HPC. On this engine the system is called Rotor Active Clearance Control System and it delivers warm air from the HPC into the HPC rotor drum for the clearance control function. In Fig. 2.11 the flow path of the cooling air into the HPC rotor is shown.

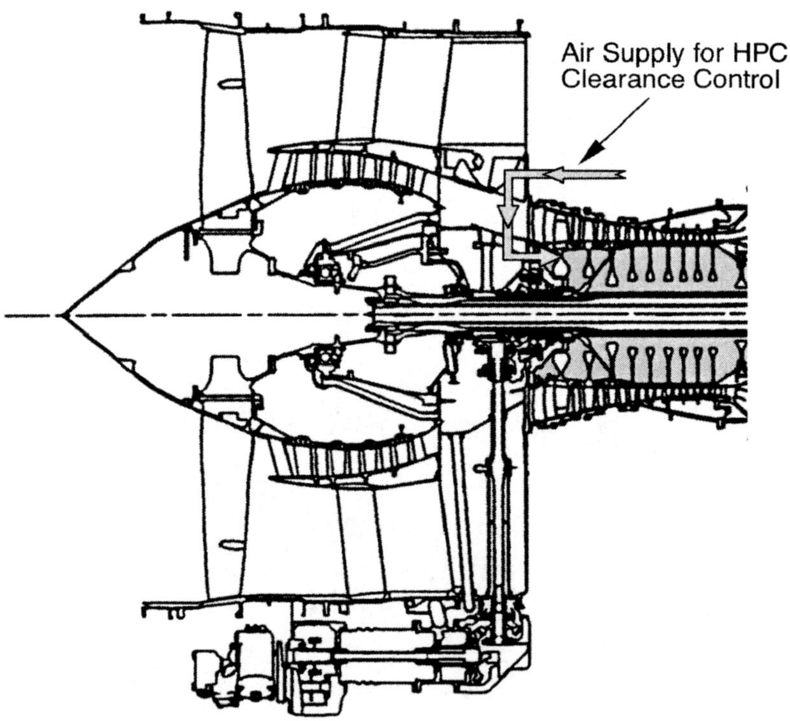

Fig. 2.11 Principle of HPC clearance control (© CFM)

2.2.2.1 Components of an Active Clearance Control System

The cooling air is extracted from the fan duct or from the appropriate stage of the HPC. In the air duct, which guides the cooling air towards the turbine case area, an airflow control valve is installed. This turbine clearance control valve is used to control the amount of air flowing to the turbine case. Because of the different cooling demands of the HPT and the LPT one valve for each turbine is needed (and one for the HPC if HPC ACC is installed). For this function modulating valves and two-position valves (RR Trent) are used. To have an even distribution around the case, the air is supplied through a cooling tube assembly installed around the turbine case. The tubes of this cooling tube assembly have tiny discharge holes on the surface facing the turbine case. Fig. 2.12 shows these tubes on the LPT case of a CFM56-7B.

A more complex design for the HPT active clearance control is used on the CFM engines. Here two air sources at the HPC are used. The HPT clearance control valves of these engines are used to control the amount of cooling air and also the mixing ratio of the two airflows. Through the mixing ratio the temperature of the cooling air can be controlled. This installation is shown in Fig. 2.12.

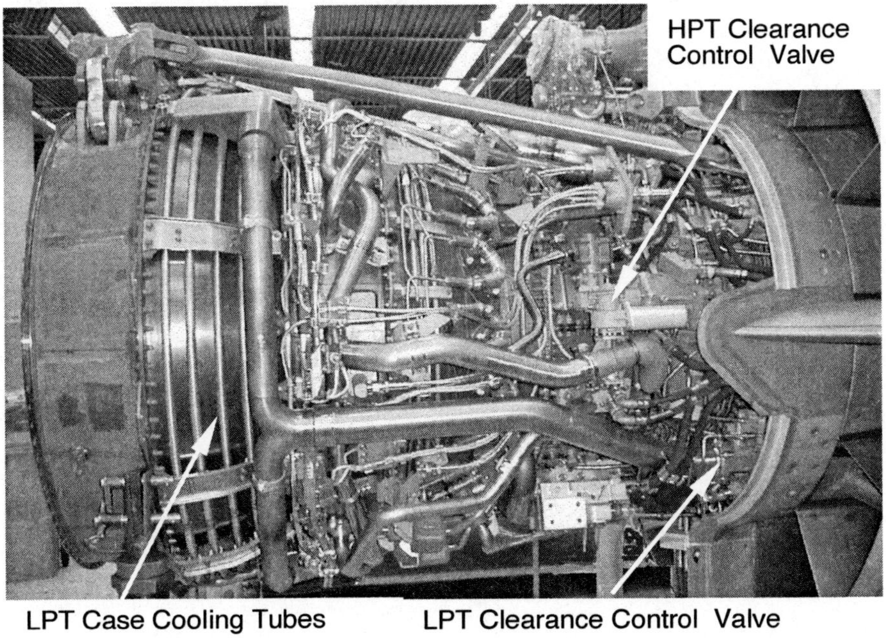

Fig. 2.12 Clearance control valves on a CFM56-7B (© LTT)

2.2.2.2 Operation of an Active Clearance Control System

As stated above, the thermal expansion characteristics of the turbine case and the turbine rotor are different. One reason for this is the different amount of material in the rotor and in the thin-walled case. The other reason is the larger diameter of the case compared to that of the rotor.

During the operation of a real engine the thermal reaction of the rotor to changes in gas temperature are superimposed by the effect of the centrifugal force. This means that the rotor expands due to heat and centrifugal force.

The changes of the tip clearance due to these effects are fastest during acceleration and deceleration of the engine. Therefore the control logic of the system must consider the dynamic and the static operating conditions.

Figure 2.13 shows how the tip clearance of a typical HPT changes during the different engine operating conditions without tip clearance control and with the influence of an active clearance control system:

1. Tip clearance during steady state operation:
   When the shaft speed stabilizes after the acceleration from a low power setting to a high power setting, the turbine case and the turbine rotor receive more heat from the gas flow compared to the lower power setting. Both parts are expanding. The rotor expands also due to the increased centrifugal force. Thus it expands more than the case. This results in a smaller tip clearance. If the higher power setting is kept up, the case slowly expands more and more. So the tip clearance becomes larger. To keep the tip clearance small, more cooling is required than at the lower power setting. The smallest tip clearances are reached during cruise flight after the power setting has stabilized for a certain time. On this condition the rotor operates at a constant temperature and the expansion has stopped.

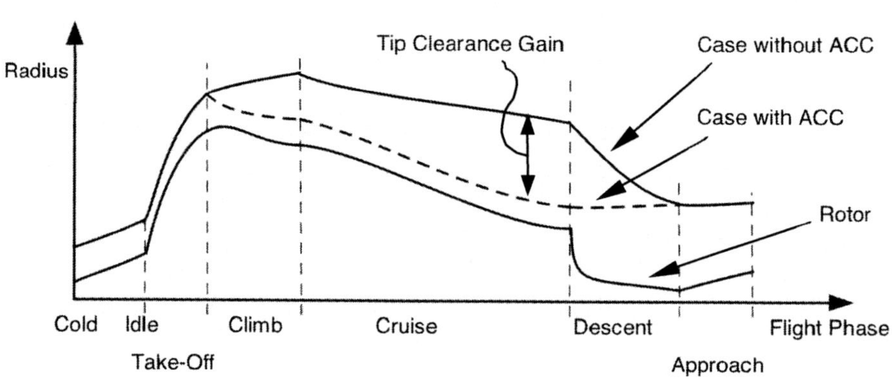

Fig. 2.13 Change of HPT tip clearance during the flight phases

This is the moment for the cooling system to increase the cooling for another reduction of the tip clearance without the danger of tip rubbing.

An exception to this is the operation with extreme gas temperatures at take-off power. Here the rotor expands with the effect of the centrifugal force and the effect of the extreme gas temperatures. To prevent the turbine from rubbing, the cooling of the case is set to a minimum or is switched off.

After the change to a lower power setting, the tip clearance changes in the opposite direction. The gas temperature is lower and the centrifugal force has decreased. Now the rotor shrinks more than the case and the tip clearance becomes larger. The cooling system will increase the cooling airflow to cool the case on the lower temperature level for a small tip clearance.

2. Tip clearance during acceleration (see also Fig. 2.14):
   During acceleration the temperature of the gas flow increases. The rotor starts to expand very fast due to heat and centrifugal forces. The tip clearance becomes smaller. Here the cooling system must react with a decreasing cooling effect. Otherwise a rubbing of the turbine rotor occurs. Later in the acceleration phase the case expands more than the rotor and the tip clearance opens again. The cooling is modulated in such a way that the tip clearance remains small during the whole acceleration and the rotor does not rub.
3. Tip clearance during deceleration (see also Fig. 2.14):
   During deceleration the heat transferred from the gas flow to the case and to the rotor decreases. So both components start to shrink. Because the centrifugal force of the rotor decreases too, the rotor shrinks

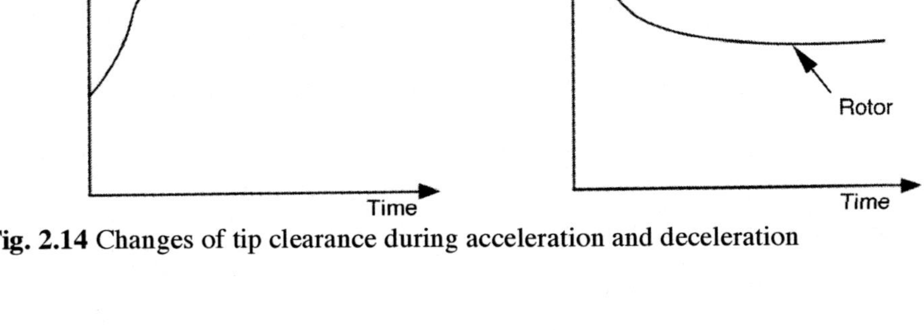

Fig. 2.14 Changes of tip clearance during acceleration and deceleration

faster than the case during the first seconds. The tip clearance opens. The cooling system must then increase the cooling effect to adapt the shrinking rate of the case to that of the rotor. At the end of a deceleration the tip clearance must be large enough to prevent a tip rub if an acceleration follows.

2.2.2.3 Control Logic and Inputs

For the operation of the clearance control system the electronic engine control (EEC) of the FADEC system calculates the position of the clearance control valve. To do this calculation the EEC needs some input values, generally the rotor speed (N1 and/or N2), the altitude and the operating condition (acceleration, deceleration, steady state). The more complex system for the HPT of the CFM56, as shown in Fig. 2.15, uses some additional inputs because the calculations done are also more complex. The inputs are then the following:

- N2
- Tcase
- TAT
- P0
- T3
- T25
- Transient bleed valve transient state signal

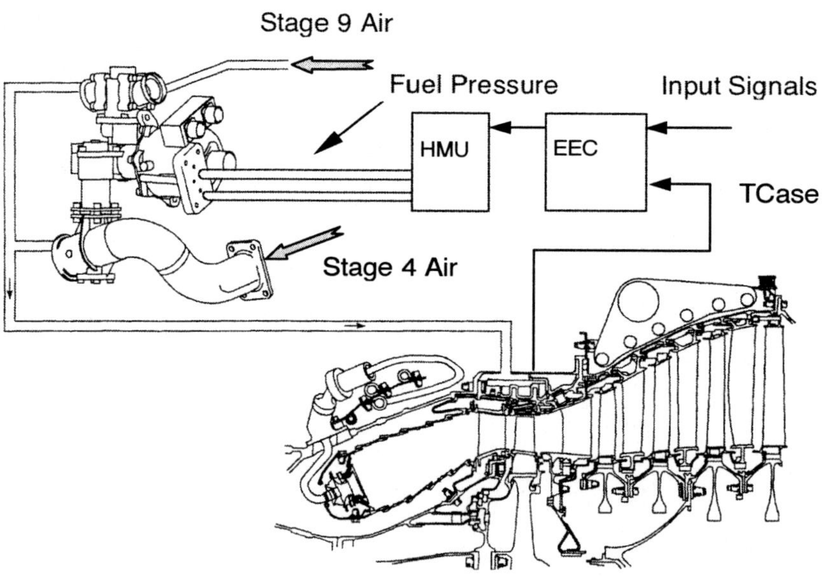

Fig.2.15 HPT clearance control system of a CFM56-7B (© CFM)

The calculation is done in these steps:

- The EEC calculates the thermal state of the HPT rotor and the air available in the primary gas flow of the engine using N2, TAT.
- It sets a theoretical demand and clearance adjustment for thermal expansion according to P0, N2, T3.
- It calculates the valve position using N2, T3, T25, Tcase and the Transient Bleed Valve signal.

To get the cooling effects described above on the systems using fan air modulating valves, the HPT cooling airflow is released only during the climb and cruise phases of a flight. On the CFM56 engines the HPT cooling operates during all operating conditions with the respective cooling effect.

On the RR Trent 700 engines with two-position valves, the valves are opened during the cruise phase only. At the higher power settings with the closed valve the case diameter is kept large because no cooling air flows. Thus the rub of the blades is prevented. This system with the two-position valve for IPT/LPT is shown in Fig 2.16.

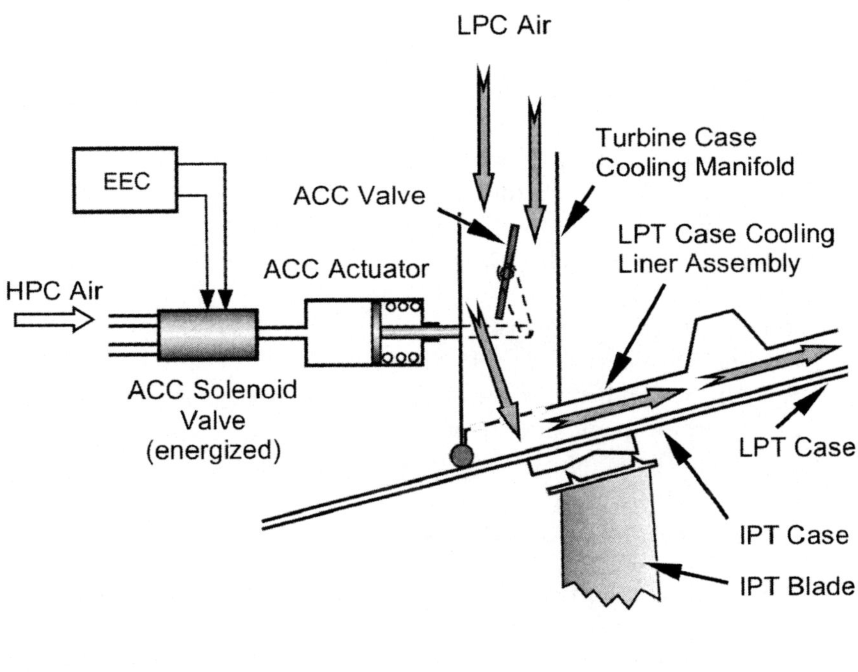

Fig. 2.16 Two-position active clearance control valve of a Trent 700

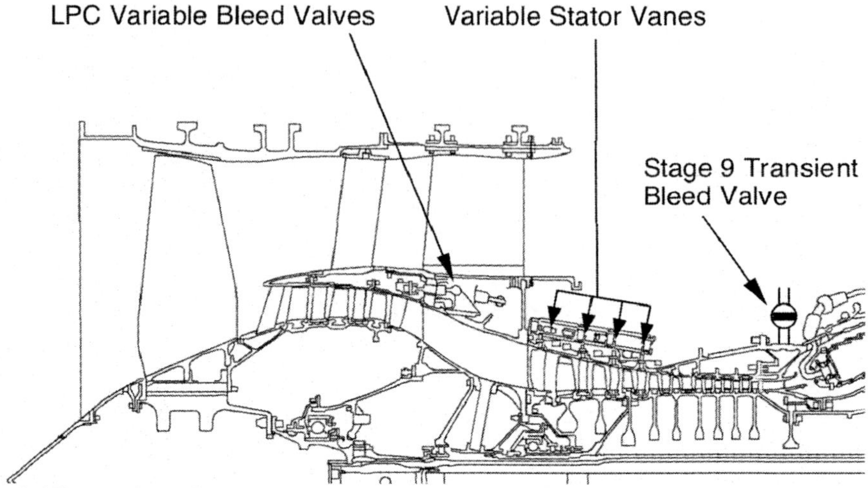

Fig. 2.17 Location of compressor control system components on a CFM56-7B (© CFM)

2.2.3 Compressor Control Systems

The compressors of a commercial turbofan engine are designed for optimum operation at the design point. This is the operating condition during cruise flight. For operation on power settings below cruise power the compressors need systems that keep the airflow within the limits and prevent stall and surge. To ensure this, compressor airflow control systems are installed. On a typical twin-spool engine, as shown in Fig. 2.17, these are:

- the variable bleed valve (VBV) system (or surge bleed system) at the exit of the LPC
- the variable stator vane (VSV) system for the HPC
- bleed valves at certain stages of the HPC

On hydromechanically controlled engines these systems are controlled by the hydromechanical fuel control unit or a separate controller. On FADEC-controlled engines they are controlled by the FADEC computer.

2.2.3.1 Variable Bleed Valve System

When the speed from the LPC decreases below the design point, the axial velocity of the airflow decreases, too. This leads to an increasing angle of attack at the rotor blades. To prevent the angle of attack from reaching a critical value the variable bleed valves are opened to bleed a portion of the airflow out of the gas path at the exit of the LPC. Because the bleeding of

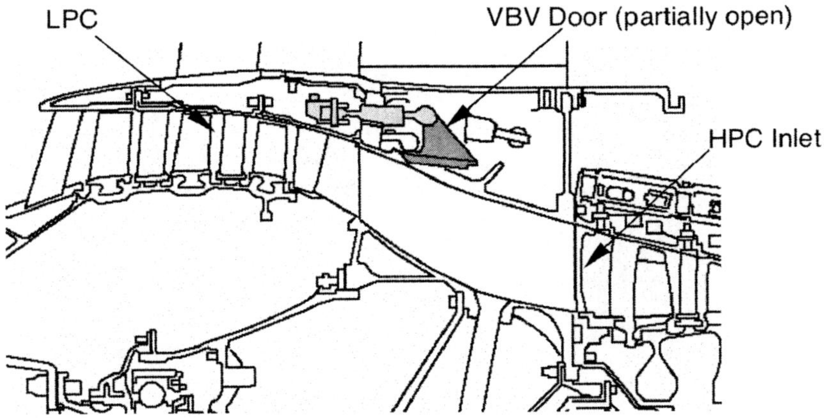

Fig. 2.18 Location of VBV doors in a CFM56-7B (©CFM)

compressed air is a loss of energy, the VBVs are opened no more than necessary to allow the specific amount of air to leave the gas path. Figure 2.18 shows a VBV door in an intermediate position. For this operation the VBVs are increasing their exit area with decreasing speed of the LPC.

Different mechanical designs for the VBVs are used. In engines from GEAE and CFM hinged doors cover openings between the struts of the engine case (fan frame) located behind the LPC. Figure 2.19 shows such a VBV mechanism installed in a CFM56-7B.

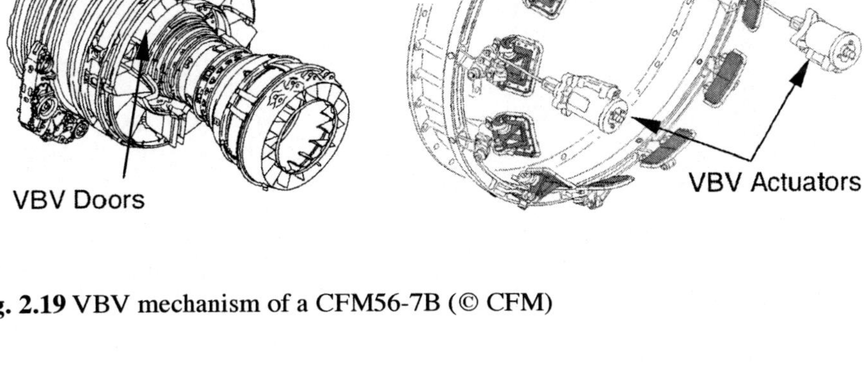

Fig. 2.19 VBV mechanism of a CFM56-7B (© CFM)

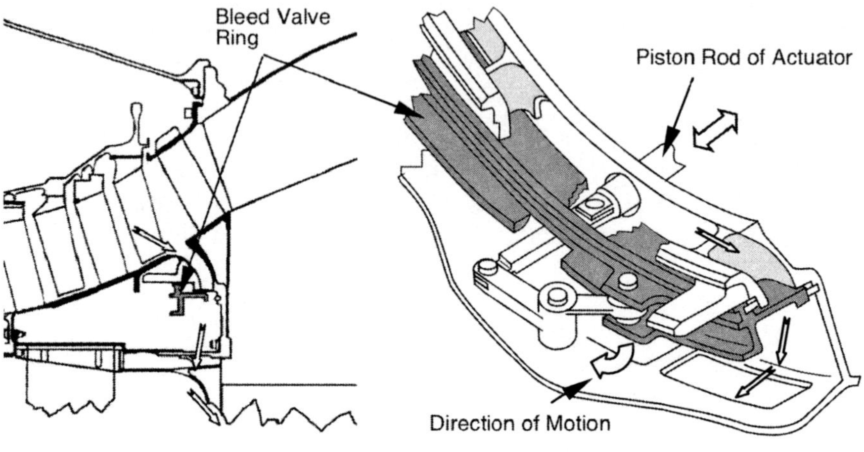

Fig. 2.20 Bleed valve ring of a PW4000 shown in closed position. Flow arrows show the bleed airflow in the open position

In engines by Pratt & Whitney and IAE axial moving concentric rings are used to cover a circumferential slot at the exit of the LPC. The system is usually operated by one or two hydraulic actuators and controlled by the EEC. The electrical position sensors for the closed loop control are installed within the actuators. Fig. 2.20 shows the bleed valve of a PW4000 and the airflow of the bleed air.

For the calculation of the position of the VBV the electronic engine control (EEC) uses several inputs. Which inputs are used depends on the design philosophy of the control system. The input parameters of different engine types are listed as follows.
PW4000 and V2500:

- N1 speed
- N2 speed
- Mach number (Mn)
- Altitude
- TRA
- T2

The primary parameter for the calculation of the VBV position is the N1. Here a direct relation exists between the N1 speed and the required position of the VBV as shown in the simplified schedule in Fig. 2.21. This calculated VBV position is biased by the use of the other parameters:

- The use of the N2 speed takes into account the change in the rotor speed matching and the operating line during acceleration and deceleration. During deceleration the surge margin of the LPC becomes smaller.

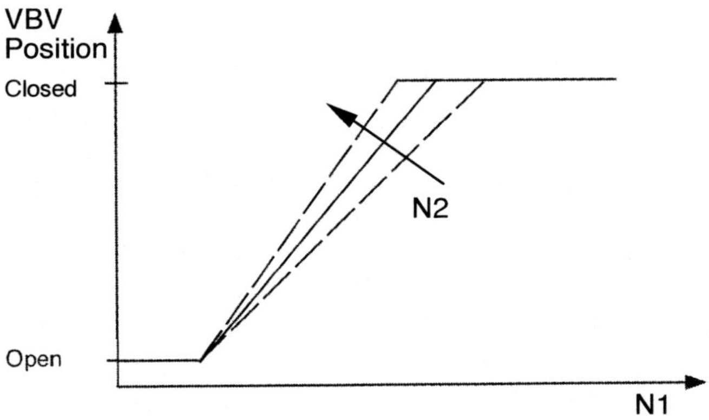

Fig. 2.21. Simplified VBV Schedule based on N1 with N2 Influence

A sufficient surge margin is retained with the VBVs more open. Thus the VBV schedule during dynamic operation is different from the schedule at steady state operation.

- With Mn, altitude and T2 the movements of the surge line are taken into consideration.

CFM56-7B:

- N2
- N1
- VSV position
- T12
- Altitude

Here the primary parameter for the calculation of the VBV position is the N2. The related schematic is shown in Fig. 2.22. This calculated position is biased by the use of other parameters:

- The influence of temperature is considered by the use of the VSV position. The VSV control system calculates the VSV position using T25.
- With the use of the N1 the change in the rotor speed matching and the operating line during acceleration and deceleration are considered here.
- With T12 and altitude the movements of the surge line are considered.

In the examples shown above the thrust lever angle is used to detect the reverse operating condition (deployed reverser) of the engine. The bleed valves are moved to a more "open" position during this special operating condition to increase the surge margin.

The VBVs are also used to assist the recovery from a stall or surge condition. If the EEC detects a stall or surge it will open the VBV to its maximum open position.

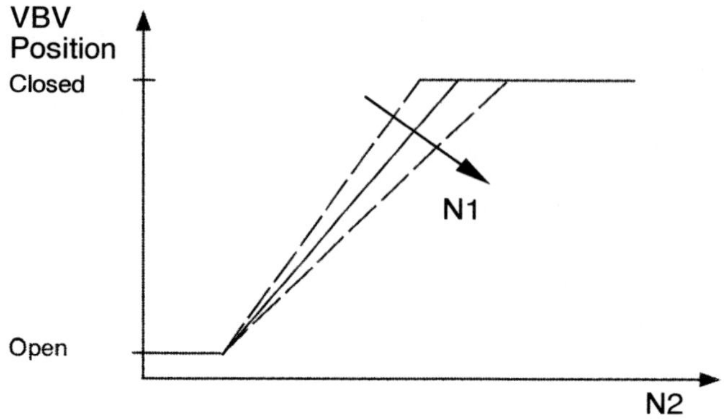

Fig. 2.22 Simplified VBV schedule based on N2 with N1 influence

2.2.3.2 Variable Stator Vane System

During low operating speeds the angle of attack at the compressor rotor blades of the front stages increases. To keep the angle off attack at its optimum during all operating speeds of the HPC the inlet guide vanes and the stators of the first 3 to 4 stages (up to 6 stages for the different CF6 models) are designed as Variable stator vanes. In a three-spool engine there are typically fewer stages equipped with VSVs than in a twin-spool engine. In the Trent 700 and 500 for example the IP compressor has variable IGVs and VSVs in the stages 1 and 2.

The mechanical designs of the variable stator vane systems of different engine types are quite similar. The turnable vanes have an inner and an outer trunnion. The inner trunnion fits into a bore of the inner shroud ring. The outer trunnion fits into a bore in the compressor case. The outer end of the outer trunnion protrudes through the case. The vane lever and the vane nut are installed on the outer end. In Fig. 2.23 the vane levers are visible. The free end of the vane lever has a trunnion that fits into a bore of the actuation ring. Each actuation ring consists of two halves bolted together. The half ring design is necessary to make the assembly of the components possible. For the movement of the vanes all actuation rings of the compressor must be turned simultaneously for a certain angle. During its movement the actuation rings are sliding over several sliding pads on the case, which provide the centering of the rings. These pads also prevent the vibration of the actuation rings together with the vane levers. The system is driven by one or two hydraulic actuators. The movement of each actuator is transferred to the actuation rings via an actuation mechanism. If one large or two small actuators are used depends on the space that is available for the actuator and its actuation mechanism.

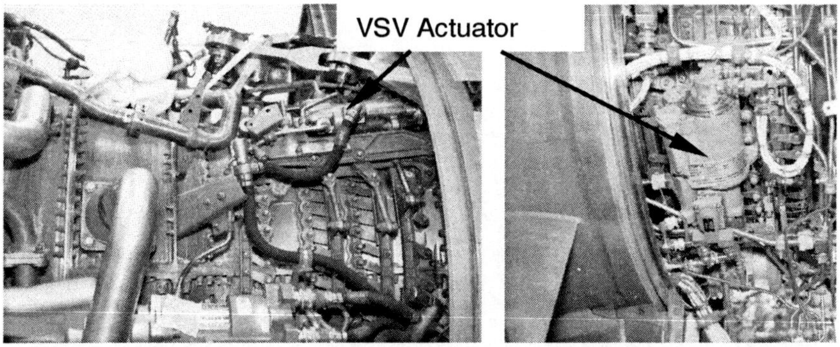

Fig. 2.23 The left photo shows the right of two VSV actuators on a CFM56-5A. The right photo shows the single VSV actuator of a V2500-A5 (© LTT)

The vane angle changes during the engine operation between the maximum low speed position (closed position) and the maximum high speed position (open position) to keep the optimum angle of attack at the rotor blades. The EEC calculates the position of the VSV actuator according to the programmed schedules and moves it to the calculated position. The required vane angle is primarily dependent on the N2 speed and corrected for the T25. So N2 and T25 are the primary parameters for the calculation of the VSV position. For a typical VSV control system of a FADEC-controlled engine the following input parameters are used by the EEC:

- Primary Parameters
  - N2, T25
- Secondary Parameters
  - Altitude
  - Throttle resolver angle TRA
  - N1
  - Total air temperature TAT

The schedules are programmed for steady state, deceleration and for acceleration. A simplified example is shown in Fig. 2.24. The two rotor speeds and the altitude are used to bias the VSV control during acceleration. An acceleration is detected by the change of the shaft speeds. The use of the acceleration schedule results in more closed positions of the VSVs to maintain a sufficient surge margin during acceleration. With increasing altitude the acceleration schedule moves further into the closed direction to keep the surge margin even in high altitudes.

During rapid accelerations and during rapid accelerations after a deceleration the vanes are moved slightly to a position that is more closed than during slow accelerations to prevent too small a surge margin under these

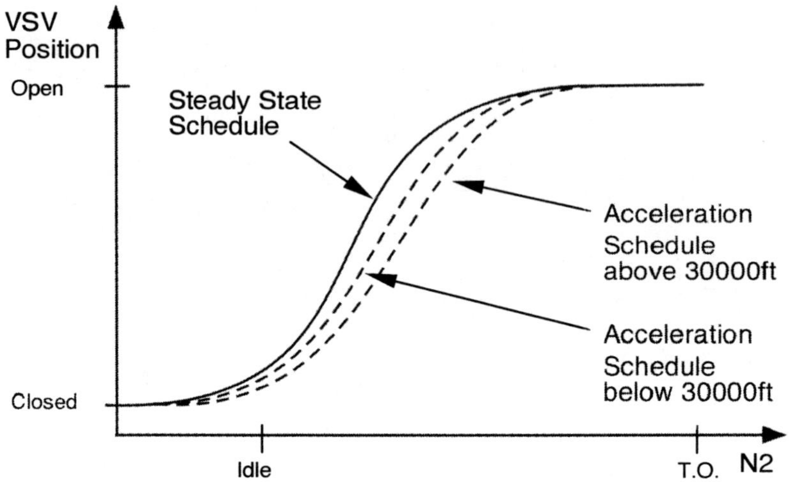

Fig. 2.24 Simplified VSV schedule with steady state schedule and acceleration schedules

conditions. The surge margin decreases during a rapid acceleration because more fuel is injected than for a slow acceleration and the back pressure from the combustion system is higher. In Fig. 2.25 the reduced surge margin is visible in the compressor map shown.

The ambient TAT is used for the detection of possible icing conditions below a specified altitude (20,000 ft for example) to protect the compressor from stalling during icing. If the given conditions are fulfilled (alt. below 20,000 ft and TAT below +2°C) the calculated VSV position is modified to a position of a few degrees more closed to increase the surge margin.

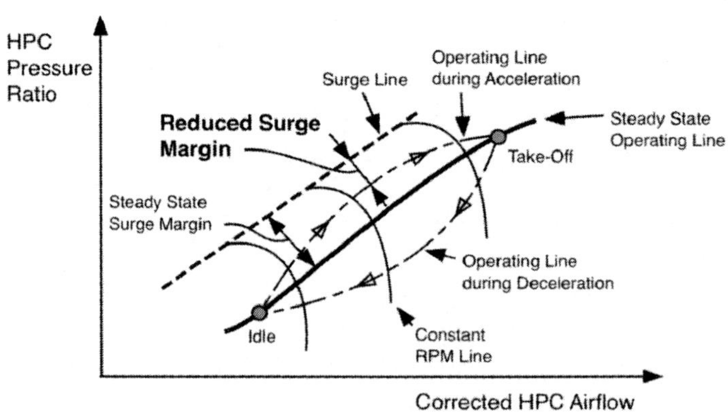

Fig. 2.25 Reduced HPC surge margin during acceleration

The VSVs are also used for a purpose that has nothing to do with the surge margin. When the acceleration schedule (or a special deceleration schedule) is used during deceleration, the VSVs are more closed than on the steady state schedule. This reduces the airflow and the time the engine needs to decelerate. Some engines that use the steady state schedule during normal deceleration use this feature only during the recovery from an overspeed condition simultaneously with the reduction in fuel flow.

2.2.3.3 HPC Bleed Valves

Additionally to the VSVs it can be necessary to use bleed valves for the HPC to prevent stalling of the HPC at low rotor speeds during operation and engine start. They are called *handling bleed valves* or *HP bleed valves*. It is also common to name them after their purpose: *Start bleed valves* or *stability bleed valves*. If bleed valves are used, the number of installed valves and the positions at the compressor stages are different from engine type to engine type. On the CF6 engines, for example, no bleed valves are used on the HPCs of all types.

The bleed valves used on HPCs are mainly two position valves (open and closed). Such valves are installed on one or two of the rear stages and opened during engine start, low idle speeds and for acceleration. They are also used for recovery from a compressor surge.

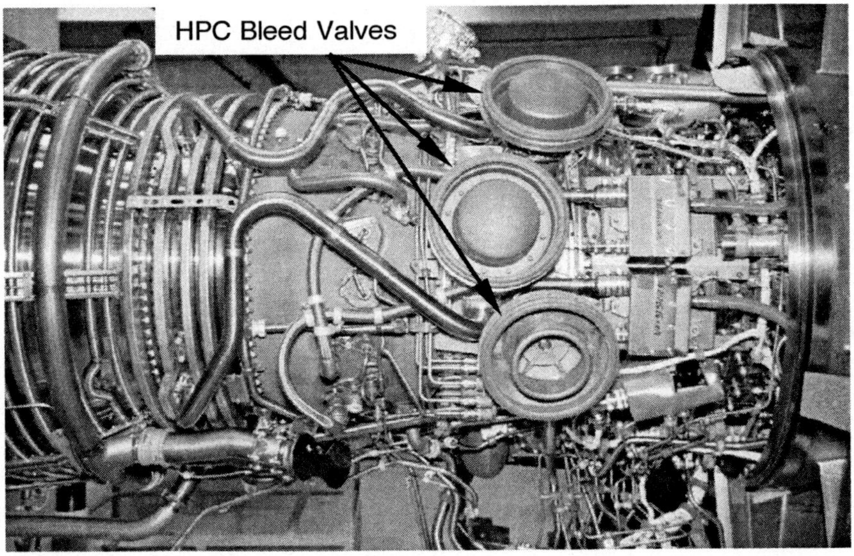

Fig. 2.26 Three of 4 HPC bleed valves of a V2500-A5 (© LTT)

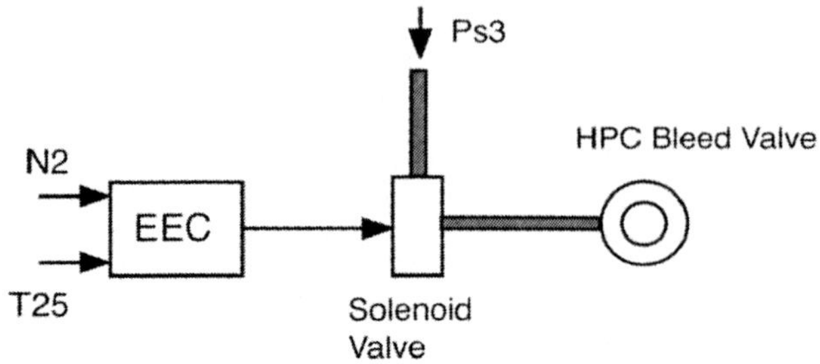

Fig. 2.27 Simplified system schematic of a pneumatically actuated HPC bleed valve. If an engine has more than one valve, one solenoid valve for each bleed valve is installed.

Here are some examples of HPC bleed valve installations:

- PW4000 2 bleed valves 5th stage
- V2500 3 bleed valves 5th stage, 1 bleed valve 7th stage
- Trent 700 4 bleed valves St. 8 IPC, 3 bleed valves St. 3 HPC

On the CFM56-7B and CFM56-5B the HP Bleed Valve is called Transient Bleed Valve and is connected to the 9th stage of the HPC. This Valve is actuated hydraulically with fuel pressure.

The bleed valves of the PW4000, the V2500 and the Trent engines are actuated pneumatically. The air pressure (Ps3) is used to open the valve. To close the valve the air pressure is vented from the valve and the compressor air pressure from the gas path forces the valves to close. The switching of the air pressure to the valve is accomplished by the use of a solenoid valve, which is controlled electrically by the EEC.

As stated earlier the HP bleed valves are used during engine start, low speeds at and above idle as well as for acceleration.

At first the operation of the single valve installation on the CFM56-7B is presented. Here the valve is opened during engine start until the engine has reached idle and the starting sequence is finished. During acceleration above idle it is opened until the acceleration is finished or the N2 speed has passed 80%. The valve position depends only on N2 speed.

Secondly the valve system operation of the V2500-A5 is described. On such an engine with a number of bleed valves all these valves are open during engine start. At N2 speeds above idle the use of the valves is different from one engine type to another.

Table 2.1 shows the schedule for one HPC bleed valve of the V2500-A5 as an example. The valve is open or will open (at decreasing N2 speeds) below the respective N2k speeds. Between the N2 value for opening and closing a hysteresis is set within the schedule. For the temperature correction the EEC uses the HPC inlet temperature of the engine.

Table 2.1 HPC bleed valve schedule of the V2500. Speeds are N2k speeds. Idle speed is 8800 RPM, max. N2 is 14950 RPM

Operating Condition	Open below	Closed above
Starting	6800	7000
Acceleration		12050
Deceleration	11600	
Surge Recovery	12352	12562

3 Engine Lubrication System

3.1 General

Within a turbofan engine the lubrication system serves several functions essential to the safe and reliable operation of the engine. These functions are:

- Lubrication of the rotor bearings
- Lubrication of the gears and bearings of the gearboxes
- Cooling of the bearings especially in the turbine area
- Removal of the contaminants from the lubricant
- Support of the sealing of the carbon bearing seals
- Supplying of a squeeze film between the bearing outer races and their housings for oil dampened bearings. Oil damping dampens the transmission of dynamic loads of the rotors to the casings. This feature reduces the vibration levels and the fatigue loads for the casings.

The lubricant reduces friction by replacing solid friction with fluid friction. Thus it must be able to provide a stable film between metal surfaces moving relative to each other at high relative velocities under high loads and high temperatures.

3.1.1 Properties of Engine Oil

Generally low viscosity synthetic lubricants and no mineral oils are used in turbine engines because synthetic oils retain their lubricating properties and are more resistant to oxidation at high temperatures. These oils also have better characteristics concerning thermal stability and the viscosity.

The oils used for turbine engines are required to operate over a wide range of temperatures. Temperatures from minus 40°C to more than 250°C for maximum bearing temperatures are possible.

Today oil of the fourth generation of synthetic oils is available. According to the development generation the oils are designated as Type 2, Type 3 or Type 4 oils. Type 2 oils are still available and in use.

The main characteristics of engine oil are:

- Viscosity
- Pour point
- Flash point
- Pressure resistance
- Oxidation resistance
- Thermal stability

3.1.1.1 Viscosity

The viscosity is the most important characteristic of engine oil. It is a measure of the internal resistance of a fluid to deform under shear stress. It is commonly perceived as "thickness" or resistance to flow. Viscosity describes a fluid's internal resistance to flow and may be thought of as a measure of fluid friction.

The viscosity of the oil depends on the temperature of the oil. It is high at low temperatures and vice versa. This means that warm oil with a low viscosity has a low internal resistance. A low internal resistance is an advantage, but if the viscosity gets too low, the load carrying capability of the oil decreases and the oil film can no longer separate the moving surfaces. Thus the lubrication capability is no longer ensured. The viscosity is usually measured in centistokes (cS). The viscosity of a typical Type 3 oil is around 5.3 centistokes at 99° C and higher than 12,000 centistokes at a temperature of –40° C.

3.1.1.2 Pour Point

The pour point of the oil is reached if it is cooled down to a temperature at which the oil becomes so thick that it stops flowing. Typical Type 3 oils for jet engines have a pour point of –62° C (-80°F). Thus, if the temperature is lower, the oil stops to flow.

3.1.1.3 Flash Point

The flash point of a flammable liquid (here engine oil) is the lowest temperature at which it can form an ignitable mixture with air. It should be as high as possible to avoid fire in the oil system. Typical Type 3 oils have a flashpoint higher than 250° C (482°F).

3.1.1.4 Pressure Resistance

The pressure resistance or load carrying capability of the oil is an important factor for the formation of an oil film between the moving parts. This film resists the loads on the moving surfaces of bearings and gears and prevents contact between the surfaces.

If the loads are higher than the pressure resistance capability of the oil, the surfaces come into contact and heavy material wear occurs. A typical value for a Type 3 oil is 2,715,114 Ryder Gear, av. lb/in % Hercolube A.

3.1.1.5 Oxidation Resistance

Oxidation is the reaction between oil and oxygen. When the oil reacts with oxygen it gets thicker and increases its viscosity due to the formation of acids and sludge. This reaction reduces the lubrication capability of the oil. The reaction rate with oxygen increases when the oil temperature and the extent of contamination increase.

Therefore the oxidation resistance is an important characteristic of oil because it influences the durability of the oil. Typical Type 3 oils are resistant to oxidation at oil temperatures up to 220° C (428°F).

3.1.1.6 Thermal Stability

The term thermal stability describes the resistance of the oil to decomposition of the oil compounds at high temperatures. The oil molecules are made of several individual compounds. At high temperatures these molecules can break apart and the chemical composition and the lubrication capability of the oil changes.

This decomposition usually occurs at very high temperatures, well above the normal operating temperatures of the engine oil. Type 3 oils can resist chemical decomposition at temperatures of up to 340° C (662°F).

3.1.2 System Design and Components

The lubrication systems generally used in commercial turbofan engines to serve above-mentioned objectives are self-contained recirculatory systems. In such systems the oil is distributed to the locations where it is needed and returned to the tank by pumps. Three subsystems are essential for the circulation of the oil. These are:

- Storage and supply system
- Scavenge system
- Venting system

The function of the storage and supply system is to deliver the required amount of oil to the bearings and gears in a condition that achieves good

lubrication. Thus the oil must be filtered and its temperature has to be in the proper range. The system uses the amount of oil stored in the oil tank. The function of the scavenge system is to return the oil from the sumps in the bearing compartments and the gearboxes to the oil tank. In many oil systems the scavenge oil passes a filter before it enters the tank. Due to the operation of the bearing compartment seals an airflow into the bearing compartments across the seals is necessary. The vent system ensures the venting of this air to the atmosphere. It keeps most of the oil droplets from the vent air within the oil system.
The major components of a typical lubrication system are (see Fig. 3.1):

- An oil tank
- An oil pressure pump and supply lines
- Scavenge pumps and return lines
- Filters and strainers
- An oil cooler
- A venting system
- Sensors for flight deck indication and condition monitoring.

Additionally servicing provisions are installed. These are the oil tank fill port and hose connections for remote filling of the oil tank.

As each area that needs lubrication has its specific requirement concerning the amount of oil for proper lubrication and cooling, the oil flow to the

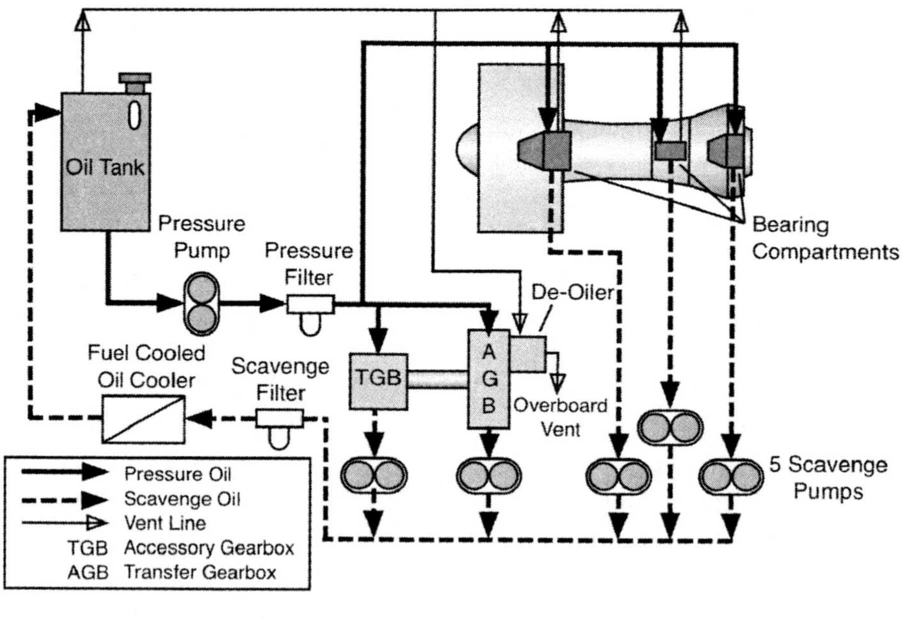

Fig. 3.1 A typical lubrication system for an engine with 3 bearing compartments

lubricated areas has to be matched with their demand. This matching is achieved by the cross section of the supply lines and the oil nozzles.

Concerning the pressure regulation, two types of systems are in use. The *relief valve system*, also called the constant pressure system, and the *full flow system*. In the relief valve system the pressure at the pump exit is maintained at a specific value over the engine operating range by a relief valve that returns excessive oil into the tank.

The full flow system operates without any pressure regulation device. Thus the oil flow in the supply lines is a function of the operating speed of the pressure pump, the supply line and oil nozzle cross sections and of the oil viscosity. This leads to a changing oil pressure with changing engine shaft speeds.

The pump of a relief valve system also supplies the part of the oil flow that is returned to the tank to keep the pressure constant. It has to be larger and requires more power for operation than the pump of a full flow system.

Most systems are designed as full flow systems. This type of system uses smaller pumps compared to the constant pressure system. Such a system saves weight and is easier to adjust, because it has no pressure regulating valve.

Independent of the method of supply oil pressure regulation the subsystems of both system types are practically identical. Figure 3.1 shows the main components of a typical lubrication system. This example shows a full flow system with a fuel cooled oil cooler. The vent (or breather) system uses a central de-oiler installed on the accessory gearbox.

3.2 Lubricated Areas

The rotor bearings used in turbofan engines are ball bearings and roller bearings. They are located within sealed bearing compartments. Oil supplied by the pressure pump is directed through individual supply lines and sprayed on the bearings through one or more nozzles per bearing. Figure 3.2 shows the arrangement of the oil nozzles for the front ball bearing of a CFM56-3. The oil-wetted areas are only inside the bearing compartments. The oil has no contact to the rotor components outside the bearing compartments and to the gas path. To ensure this the walls of the bearing compartments are sealed against the rotating shafts. For this sealing two types of seals are used: The carbon seal and the labyrinth seal. To have a constant sealing efficiency an airflow exists across the seals into the bearing compartments as shown in Fig. 3.3.

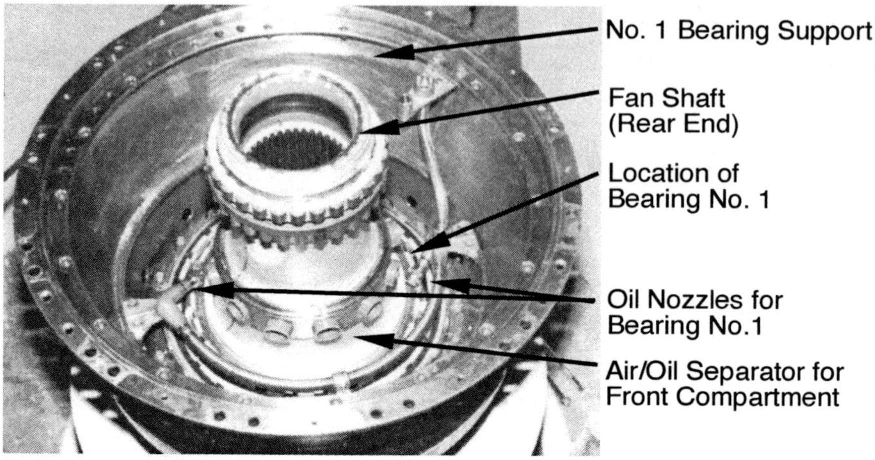

Fig. 3.2 The view from the rear into the front half of the front bearing compartment of a CFM56-3 after its removal from the engine. The oil nozzles for Bearing No.1 and its supply line are visible (© LTT)

In the gearboxes ball and roller bearings are used to support the gear shafts. The gears and the bearings are supplied with oil through oil nozzles like the engine rotor bearings. The oil that drips off the bearings and gears is collected on the bottom of the bearing compartments and the gearboxes. These areas are called the sumps. Through return lines scavenge pumps take the oil out of the sumps with a higher capacity than the pressure pump delivers the fresh oil into the bearing compartments or gearbox. Thus the scavenge pump prevents the accumulation of oil in the sumps. Due to the

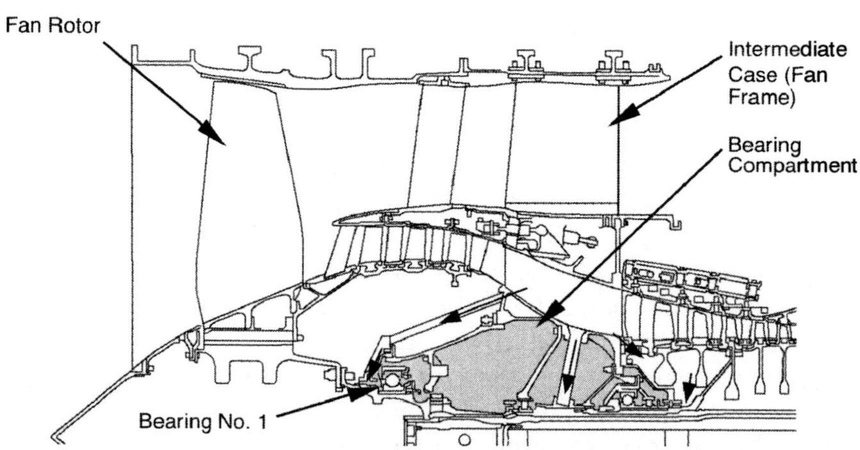

Fig. 3.3 The front bearing compartment of a CFM56-7B with the 3 forward bearings. The sealing airflow into the compartment is shown (© CFM)

higher capacity of the scavenge pump it returns oil with a substantial amount of sealing air from each sump to the oil tank. The return lines and scavenge pumps must be properly sized and designed to handle the oil/air mixture. This generally requires an individual scavenge pump for each sump and the venting of the oil tank.

3.3 System Components

3.3.1 Oil Tanks

The oil of the lubrication system is stored in a tank. To allow servicing, devices for filling and draining the oil tank are provided. For the check of the oil level a sight glass or dipstick is installed. For the remote indication of the oil level an electrical quantity sensor is located in the oil tank.

The recirculatory system generally provides two different locations for the oil coolers in the system. If the oil is returned from the sumps directly to the tank without cooling, the system is called a *hot tank system*. If the oil passes the oil cooler before it enters the oil tank, the system is called a *cold tank system*.

Typical locations for the oil tanks on the engine are either the fan case or the accessory gearbox. If the engine has a core engine-mounted accessory gearbox with an oil tank, the access to the oil tank for servicing is not as good as the access to a fan case-mounted oil tank. The oil tank of the PW4000 for example requires an access door in the inner barrel of the thrust reverser structure because this oil tank is mounted on the gearbox of this engine. This variant is possible on engines with short duct nacelles only. A fan case-mounted oil tank can be used on every engine design.

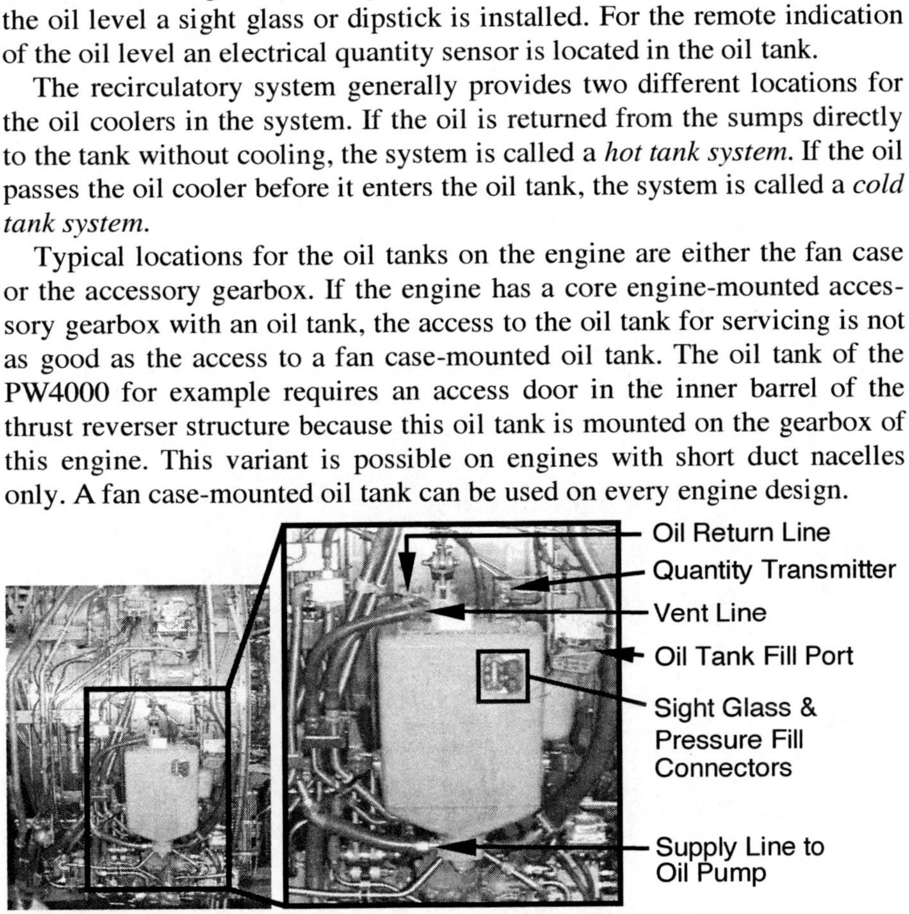

Fig. 3.4 The oil tank of a CFM56-5C on the left side of the fan case (© LTT)

The typical oil tank has three connections to the lubrication system. These are the oil supply line to the pressure oil pump, the oil return line from the scavenge pumps and the vent line. The scavenge pumps deliver a scavenge oil/air mixture into the tank. This air is vented through a static de-oiler within the tank to the de-oiler or into one of the bearing compartments. On some oil tanks a pressurization valve is installed in the vent line connection. This valve keeps an air pressure slightly above the ambient pressure in the tank after engine shutdown. This facilitates the oil supply to the pressure pump during engine start.

For the remote sensing of the oil quantity a sensor is installed in the oil tank and connected to the assigned computer.

The oil tank can be filled directly with oil through the fill port or via remote fill connections. To check the oil level a sight glass is installed in the oil tank wall. If the pressure fill ports are used for refilling the oil tank, the oil is pumped into the tank until it is visible in the overflow hose.

The capacity of oil tanks depends mainly on the engine size. Here are three examples for oil tank capacities:

CFM56-7B: 23 US quarts, 22 litres
PW4000 (94 inch): 37 US quarts, 35 litres
GE90: 28 US quarts, 30 litres

3.3.2 Pumps and Filters

The oil pumps used in lubrication systems are gear type pumps or vane pumps. The gear type pumps are of the classic type with parallel shafts or they are of the gerotor type with coaxial gears. Usually one pressure pump

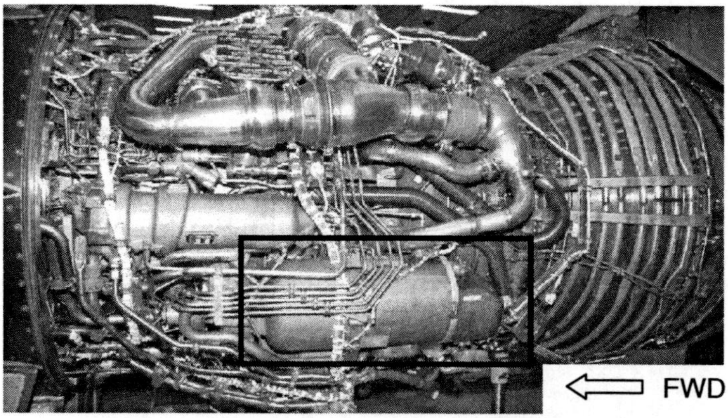

Fig. 3.5 The oil tank of a PW4000 on the core engine-mounted accessory gearbox. All connections to the lubrication system are integrated into the gearbox (© LTT)

and for each sump one scavenge pump are used in a lubrication system. This leads to a number of four to six scavenge pumps. The pumps are either arranged in separate units – one for the pressure pump and the other for the scavenge pumps, or all the necessary pumps for a system are integrated into one unit. They are installed on the assigned pads of the accessory gearbox. On engines from GE and CFM the oil pumps are integrated together with the oil filters in one unit. Here these units are called *lubrication units*. Figure 3.6 shows the installation on a CFM56-5A.

The magnetic chip detectors for debris monitoring are installed in the scavenge pump inlets or in the scavenge oil lines upstream of the pumps where easy access is ensured.

The filter system is a very important element for the reliability of a recirculatory lubrication system. Because the oil has to pass through small holes and passages, even very small particles contaminating the oil could block the oil flow resulting in a lubrication failure. The normal contaminant is abrasive material and is released by the bearings and gears during their normal operation. It is flushed away from the bearings and gears by the oil and carried with the scavenge oilflow away from the sump. In the filters of the system the contaminants are removed nearly completely from the oil. Thus the oil can be supplied again to the bearings and gears. If a bearing or gear failure develops, larger than normal particles will be found in the filters and on the *magnetic chip detectors*.

A classic filter arrangement in a lubrication system is one pressure filter downstream of the pressure pump and one scavenge filter downstream of the scavenge pumps. In this arrangement the scavenge filter has the finest

Fig. 3.6 The lubrication unit of a CFM56-5A. It contains one pressure pump and 4 scavenge pumps. The Pumps are gerotor pumps (© LTT)

filter element. Also in use are systems with one filter in the pressure system only (e.g.:PW4000, CFM56-5B, -5C). In these systems a back-up filter is installed to ensure oil filtering if the main filter is clogged. Every oil filter is equipped with a bypass valve to sustain the oil flow while the filter element is clogged. The filter elements used in the oil filters have mesh sizes between 15 and 65 Microns (0.015 to 0.065 mm).

The finest filter in the system is monitored with a differential pressure switch for clogging. The resulting clogging warning informs the flight crew about the limited filtering efficiency in the system.

In some systems additional filter screens are installed upstream of the oil nozzles in the supply lines. These screens are called last chance screens (or last chance filters). They prevent a clogging of the oil nozzles by particles that can reach the nozzles if the filter bypass valve is open.

3.3.3 Oil Cooling

During the lubrication process heat is transferred from the engine components to the oil. It has subsequently to be removed from the oil to keep the oil temperature within the set limits. This requires the installation of an oil cooler in the system. The cooling medium may be fuel or air and in some designs a combination of a fuel-cooled and an air-cooled heat exchanger is used. The cooler may be located either on the feed side or the return side of the lubrication system resulting either in a hot tank system or a cold tank system.

The typical oil cooler used on turbofan engines is the fuel-oil heat exchanger. It has a smaller volume compared to an air-oil heat exchanger of the same cooling capacity and the use of fuel as the cooling medium results in a heating of the cold fuel delivered by the aircraft fuel system. The fuel-oil heat exchanger is also effective during ground operation and need not be exposed to the airflow. The lubrication systems on some engines use additional air-cooled oil coolers to control the temperatures of oil and fuel during the operation with low fuel flows (see also Chapter 4.4).

In some fuel-cooled oil coolers a thermostatic bypass valve is installed. It maintains a proper oil temperature by varying the portion of the oil passing, respectively bypassing, the oil cooler. This valve allows changes to the cooling effect in response to the changes in fuel flow in different phases of engine operation. All other oil cooler designs have a bypass valve responding to the differential pressure across the cooler. The highest differential pressure develops with the coldest oil. Thus the cooler is bypassed when the oil temperature is low.

3.3.4 Vent System Components

In order to maintain a continuous flow of air from outside the bearing compartment through the shaft seal into the bearing compartment, the sealing air has to be vented through vent lines out of the bearing compartment. When this *vent air* or *breather air* leaves the bearing compartment, it contains a large amount of oil droplets. Furthermore the sealing air mixes with the scavenge oil, which is pumped through the return lines into the oil tank. Thus means must be provided to separate the oil from the air to retain the oil within the system and vent almost clean air into the atmosphere.

After the vent air has left the lubrication system it flows through a *de-oiler* (it is also called *de-aerator*). This is a centrifugal device for the separation of oil and air. The vent air of the oil tank also flows through the de-oiler to the atmosphere. Figure 3.7 shows the de-oiler and its vent air outlet on the gearbox of a V2500-A5.

Within the de-oiler the vent air flows through radial passages in the rotor towards the center of the shaft. Most of the oil droplets are thrown out of the rotor and against the walls of the housing by the centrifugal force. At the bottom of the de-oiler housing the oil droplets are collected and returned to the oil tank via the scavenge pump. The air from the center of the shaft is vented into the atmosphere via the vent air outlet. A small amount of oil is released with the air leaving the system. It represents the normal oil consumption of the engine (0.1 to 0.5 US quarts/hour).

When the vent air leaves the bearing compartments through the engine shaft (in engines of CFM and GE) to the rear, the de-oiler is called *air/oil separator*. In fig. 3.2 the radial tubes of the air/oil separator are visible. No vent air lines to a central de-oiler are necessary when an air/oil separator in each bearing compartment is used.

Fig. 3.7 The de-oiler of a V2500-A5 on the front side of the accessory gearbox (© LTT)

3.4 System Indications and Monitoring

3.4.1 Indication of Operating Data

The indications of a typical lubrication system show the pilots the main operating data of the system for monitoring purposes. According to EASA CS 25.1305 and FAR 25.1305 the following data must be indicated:

- Oil Quantity
- Oil Pressure
- Oil Temperature
- Low Oil Pressure Warning
- Oil Filter Clogging

The indication system provides also a warning of high oil temperature and of a low oil level in the oil tank. The sensors for these data are located at assigned locations within the lubrication system. Figure 3.8 shows the oil system indications on the display screen of an A330.

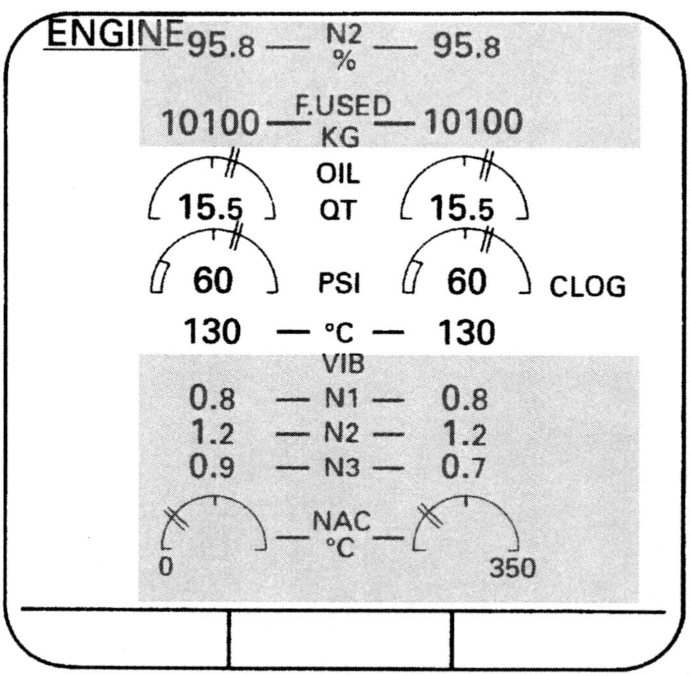

Fig. 3.8 The lubrication system indications on the ECAM display of an A330. The word *clog* indicates a clogged filter in the respective system and appears only if the respective filter clogging switch has signaled the clogged condition (© LTT)

The following table shows the lubrication system indication data of 2 engines as an example. The reason for the higher oil pressure during cruise of the V2500 is the large decrease in the vent pressure connected to the pressure transmitter. The change of the vent pressure from take-off to cruise is larger on the V2500 than on the PW4000:

Table 3.1 Indication data of the PW4000 and the V2500-A5

	PW4000	V2500-A5
Min Oil Press.	4.8 bar (70 psi)	4.1 bar (60 psi)
Oil Press. at Idle	6.2 to 8.6 bar (90 to 125 psi)	7.9 bar (115 psi)
Oil Press. at Cruise	14.5 to 17.2 bar (210 to 250 psi)	17.9 bar (260 psi)
Oil Press. at T.O.	17.9 to 19 bar (260 to 275 psi)	16.6 bar (240 psi)
Max. Oil Temp Cont.	163 °C	156 °C
Max. Oil Temp. Transient	177 °C	165 °C
Typ. Cruise Oil Temp.	120 to 125 °C	110 to 120 °C

3.4.2 Indication System Sensors

The indication system sensors are located at specific locations within the lubrication system. Its output is not used for indication only. The sensor output data is also continuously collected for system monitoring. Additionally to the indication sensors, the chip detectors for debris monitoring purposes are installed in the scavenge system.

On all engines the oil quantity sensor is located in the oil tank. It is installed into the tank from the top and mounted to the ceiling of the tank. So it can be replaced without draining the oil tank. For the oil quantity measurement sensors of the capacitance type and the reed switch type are used.

The oil pressure transmitter is connected to an oil line of the supply system and to the vent pressure of the lubrication system. If the vent pressure level is very different between the bearing compartments, the vent pressure of the bearing compartment with the highest pressure is used. Due to this way of connecting the pressure sensor to the system, the indicated value shows the pressure difference between the absolute oil pressure and the absolute vent pressure.

The oil temperature sensor can be located in the scavenge system or in the supply system. The position in the individual system is selected by the designer to keep the maximum indicated values at approximately 150°C.

For the triggering of the low oil pressure warning a low oil pressure switch is installed parallel to the oil pressure transmitter, or the dedicated computer triggers the warning based on the oil pressure sensed with the oil pressure transmitter. In the latter configuration the threshold value for the low oil pressure warning can increase with the engine shaft speed.

3.4.3 Data Processing, System Monitoring

The sensors of the indication system are connected to computers located within the airframe or to the FADEC computer. Which variant is used depends on the computing power of the FADEC computer. If the sensors are connected to the aircraft computers, these are mainly the interface computers, which also handle the data transmissions between the FADEC computer and the airframe systems.

Besides the collection of sensor data for flight deck indication there is also a collection of data that is later evaluated on the ground only to detect the development of a trend. A very common example is the monitoring of the oil consumption of an engine.

The monitoring of the oil consumption is done to sample information about the efficiency of the bearing compartment seals and for the early detection of leaks. This is achieved by using the oil quantity data delivered in regular intervals by the system. When the bearing compartment seals suffer from increased wear, the amount of sealing air flowing across the seals increases. In an early state this does not result in a leak at the affected seal. The resulting effect is a higher flow of vent air through the de-oiler. This reduces the efficiency of the de-oiler and increases the oil consumption. If the wear of the seals progresses, leakages may occur.

The oil leaking out of a bearing compartment may enter the gas path of the engine. This leads to an oil contamination of the compressor airfoils and to an oil contamination of the bleed air delivered to the aircraft. An oil smell in the cabin will be the result. Because this is not acceptable, an operation with a leaking bearing compartment must be prevented. Oil consumption monitoring is an effective tool to achieve this.

3.4.4 Debris Monitoring

Fresh oil filled into the lubrication system will gradually dissolve contaminants and holds microscopic particles in suspension. Bearings, seals and gears wear, erode and corrode introducing traces of these components into the oil flow. Thus the condition of the oil, which has been circulated in the

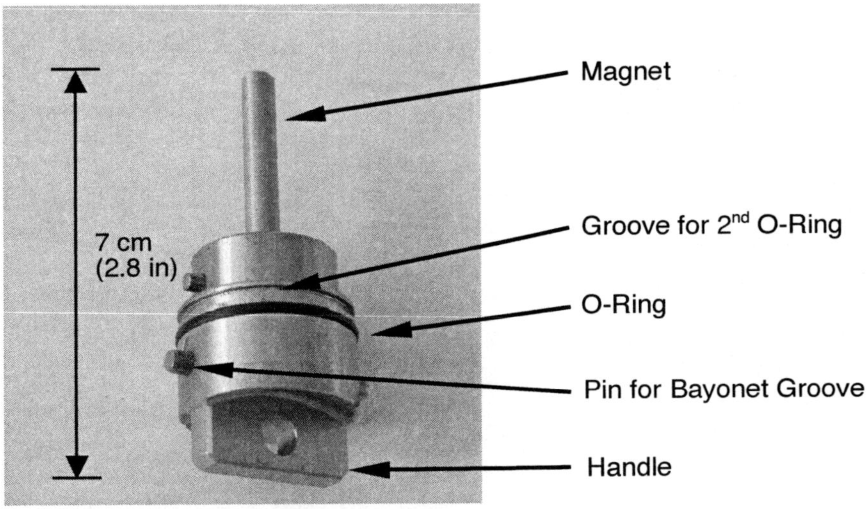

Fig. 3.9 A Magnetic chip detector for the installation in a lubrication unit. The O-rings are replaced with new ones before each installation

lubrication system for some time, very exactly reflects the condition of the system. If the system operates normally, the oil contains the amount of particles, which is typical of the system. The size of the particles is typical of abrasive contamination. During the development of a bearing or gear damage the size and the amount of the particles become larger. With the detection of these particles in the oil a bearing or gear damage can be recognized in its early stage. These particles are also collected by the oil filter, but the filter inspection intervals are too long to detect a damage early. To facilitate the check of the oil system for particles in shorter intervals, magnetic chip detectors are installed in the scavenge oil flow of each sump or as a master chip detector in the common scavenge line downstream of the scavenge pumps. Because the gears and bearings are made of steel, the magnets of the chip detectors are able to collect the particles (or chips) of these parts. These chip detectors collect particles from 0.02 to 1 mm in size. The whole arrangement of magnetic chip detectors is often called *debris monitoring system*. In its simplest design the magnetic chip detector is a tiny bar magnet which protrudes into the scavenge oil flow. Figure 3.9 shows such a chip detector. The check of the chip detectors for collected particles in fixed time intervals is part of the maintenance checks. To facilitate the check of these detectors they can be removed from their housing without a tool. If particles are found on the chip detector, they can be analyzed in a laboratory to exactly determine the component releasing these particles into the oil.

Fig. 3.10 The electrically monitored chip detector of a CFM56-5C. It is installed in the scavenge oil outlet of the lubrication unit (© LTT)

The more sophisticated variant of the magnetic chip detector is the electrically monitored chip detector as shown in Figure 3.10. Such a chip detector comprises a set of two magnets. The FADEC computer monitors the resistance between these two magnets. A check and removal of the electrical monitored chip detector is not necessary until the FADEC computer sends the corresponding maintenance message.

For the particle monitoring of modern engines *oil debris monitors* (ODM) are used. These sensors are based on an inductive measurement technique which enables the system to detect, count and classify wear metal particles by size and type (ferromagnetic or non-ferromagnetic). This allows the system to determine the trend for the amount of particles in the oil. The ODMs are connected to the FADEC computer or another computer assigned to this function.

Another tool for the monitoring of the oil-wetted parts is the *spectrographic oil analysis program (SOAP)*. Through this analysis the concentration of particles of the size from 0.001 to 0.02 mm and its specific elements can be identified. The metal type and concentration may indicate to the analyst and engineers which part of an engine is failing if a direct assignment to an engine part is possible. SOAP is used if this program is part of the engine maintenance schedule. It may be used temporarily if uncertainties exist about the reliability of engine bearings or gears. Some turbine engine manufacturers mandate SOAP in certain turbine engine maintenance schedules.

For the monitoring of particle concentration in the oil a periodic analysis of oil samples, taken from the engine, is made in a laboratory. When the amount of particles for an element or an element combination increases, this indicates an increase in wear. If the trend continues, the development of a damage is imminent. In this phase the particle size also increases and the presence of the particles can be verified by the magnetic chip detectors. The verification with the help of a chip detector is important if an assignment of the analyzed elements to an engine part is not possible.

If SOAP is used, the airline engineering has more time for the observation of a failure development. This allows a longer planning period for the engine removal necessary in such case.

4 Engine Fuel Distribution System

4.1 General

The description of the engine fuel system is divided into two parts. This chapter contains the first part and describes the engine-mounted fuel distribution system based on engines controlled by full authority digital engine control systems (*FADEC*). The second part in Chapter 5 describes the engine and fuel control system.

4.1.1 Properties of Fuel

A fuel contains chemical energy that is released as heat energy in a combustion process. The turbine engine converts this energy into thrust. Thus the turbine engine operation and the design of the fuel delivery system is largely determined by the physical properties of the fuel.

In commercial aviation two grades of turbine fuel are used: Jet A-1 and Jet A, which are both kerosene type fuels. Another grade of jet fuel is Jet B. This is a blend of gasoline and kerosene and is rarely used except in extremely cold climates.

Jet A-1 is produced according to internationally agreed standards and is generally available outside the USA. It has a freezing point maximum of -47°C.

Jet A is produced according to an ASTM (American Society for Testing and Materials) specification and is normally available in the USA. The main difference to Jet A-1 is a higher freezing point maximum of -40°C.

Jet A and Jet A-1 have a minimum flash point of 38°C. Thus the fuel must be heated prior to combustion at least up to this temperature before sufficient fuel vapor is available for ignition. In the gas turbine engine combustion occurs at a constant pressure, therefore the peak pressures occurring in piston engines are not present here. This allows the use of low octane fuel.

4.1.1.1 Freezing Point

The freezing point is defined as the temperature at which the last wax crystal melts which previously has been cooled to a temperature at which wax crystals develop.

In order to secure an uninterrupted flow of fuel from the aircraft tanks to the engines and to maintain its property as a lubricant the fuel temperature in the aircraft tanks must be monitored and maintained above the freezing point. Thus especially long range aircraft (e.g. A340) provide means of returning heated fuel from the engines to the supply tanks. In addition the fuel is also heated for other reasons before entering a fuel filter.

4.1.1.2 Viscosity

Viscosity is the resistance of a fluid to flow. Jet fuel possesses a certain degree of viscosity, which is taken advantage of by using it as a lubricant. Fuel is used to lubricate (and cool) moving parts in fuel pumps, valves, flow metering units and actuators. Because a constant flow of fresh lubricant (and coolant) is available, this allows to design near maintenance free components.

A disadvantage of the viscous properties of the fuel is that contaminants may be kept in suspension for an extended time. Thus the fuel must be filtered before it enters the engine.

4.1.1.3 Volatility

Volatility is the tendency to vaporize. This property of the fuel is of importance in several aspects. Fuel with a high volatility has the desirable property of supporting an engine start in a cold climate or an in-flight restart at high altitude. But high volatility is also the reason why fuel evaporates during storage. Thus, for practical reasons, volatility of the fuel has to be kept under a certain limit.

Fuel pumps create a suction at their inlet side. If the pressure is so low that the fluid vaporizes, the pump will draw in vapor rather than liquid thus interrupting the fuel flow (*vapor lock*). If only a certain amount of vapor bubbles is present in the liquid fuel, they will return into its liquid stage as the pressure in the pump rises. This is called *cavitation* and damages the pump components.

4.1.1.4 Cleanliness

The cleanliness of the fuel, i.e. the absence of numerous forms of contamination, is a prerequisite to incorporate components built to a high degree of precision and efficiency into the fuel system. The viscosity and a relative density close to that of water gives the fuel the tendency to hold dispersed contaminants in suspension. This makes their detection and elimination dificult. It is important that particles like dirt, corrosion products etc. are retained in filters and do not enter the fuel metering and distribution system.

Jet fuel has a tendency to absorb water. Water not only does not burn but also will freeze at the low temperatures encountered during flights at high altitude. The resulting ice crystals may plug fuel filters. Thus the fuel must be heated before entering a fuel filter. Furthermore water facilitates corrosion and the growth of microorganisms. This is the reason why the water content is limited by the production standards.

4.1.2 Fuel Supply of the Engine

In the fuel supply of an engine two systems are involved. These are the aircraft fuel supply and storage system and the engine fuel distribution system. The aircraft fuel system is also called the primary fuel system and has the functions of fuel storage as well as the supply of all engines with low pressure fuel. The engine mounted system is called the secondary fuel system. It is responsible for the fuel supply of the individual engine including the metering of the fuel for combustion.

According to its function (fuel supply of the engine driven HP pump) the whole fuel system from the tanks to the engine driven high pressure fuel pump can be seen as the low pressure fuel system.

On the aircraft side usually an isolation valve is incorporated in the fuel supply line of each engine. It is also called *low pressure shut-off valve* or *spar valve*. It allows to shut-off the fuel supply in case of an engine fire. The shut-off valve in the high pressure fuel system of the engine is used to shut down the engine. The low pressure shut-off valve is closed simultaneously with the high pressure shut-off valve during engine shutdown on most aircraft designs.

The design of the low pressure fuel system has to take into account the above-mentioned characteristics of the fuel and it has to be very reliable. Generally the system contains pumps at both ends of the low pressure supply line. Electrically driven pumps (named *boost pumps, auxiliary pumps or transfer pumps*) are installed in the tanks and an engine driven pump (often called *main fuel pump*) is installed on the engine gearbox. For each engine fuel supply line two electrically driven pumps are installed in the associated tank for redundancy. These operate either in a main pump / auxiliary pump or a main pump / standby pump configuration. Preferably these pumps are powered by brushless AC-motors, each powered by a different electric source. Brushless AC-motors are maintenance free over extended periods and thus allow installations in less accessible locations.

The low pressure system must transfer the fuel from the tank to the engine at a suitable pressure, rate of flow and temperature even at periods of peak demand without any danger of vapor locking and cavitation. Thus the boost pumps of the system are considerably oversized for normal operation. One boost pump of the A340 is able to supply all four engines (via cross feed lines) during cruise flight.

The low pressure pump of the engine is usually built in one unit with the high pressure fuel pump of the respective engine. Both pumps are driven by the same gearbox drive shaft. The main function of the engine low pressure pump is to provide sufficient pressure on the suction side of the high pressure pump so as to avoid cavitation and vapor locking. But it also provides additional redundancy in case the boost pumps have failed. A wing-mounted engine can also be gravity fed if the boost pumps and low pressure pump have failed.

4.2 Engine-Mounted Fuel Distribution System

The engine-mounted fuel distribution system delivers clean pressurized fuel for combustion and hydraulic purposes. The system ensures that the fuel has the proper temperature and pressure for its use. The fuel for combustion is metered by a metering device. This is controlled by the computer of the FADEC system, the electronic engine control (EEC). In older hydromechanical systems the metering device is a component of the hydromechanical fuel control unit (FCU).

4.2.1 System Design and Components

The basic fuel distribution system of a typical turbofan engine consists of the following components (listed in the sequence in which they are passed by the fuel):

- Fuel pump
- Fuel-cooled oil cooler
- Fuel filter
- Fuel metering device (HMU or FMU)
- Fuel flow transmitter
- Fuel manifold components
- Fuel nozzles

Figure 4.1 shows the schematic of the fuel distribution system of the CFM56-7B as a sample of a typical system. This system has all the listed components added by the fuel-cooled oil cooler for the integrated drive generator (IDG) and the servo fuel heater. Because the coolers are fuel-cooled, they are integrated into the fuel system.

The fuel pump is driven by the accessory gearbox and thus it is mounted on the gearbox housing. Here the most frequently used design is a two stage pump with a centrifugal low pressure stage and a gear type high pressure stage.

For the cooling of the engine oil the fuel-cooled oil cooler is located in the system downstream of the low pressure stage of the fuel pump and the IDG oil cooler.

To prevent small dirt particles from entering the components of the fuel systems where they can be dangerous for the operation of valves and fuel nozzles, the fuel filter is located downstream of the oil cooler. From the fuel filter the fuel flows into the high pressure stage of the fuel pump. At the exit of the high pressure stage or in the following fuel metering device the flow of the hydraulic fuel is extracted from the main flow. It is used as the hydraulic medium for the hydraulically operated actuators on the

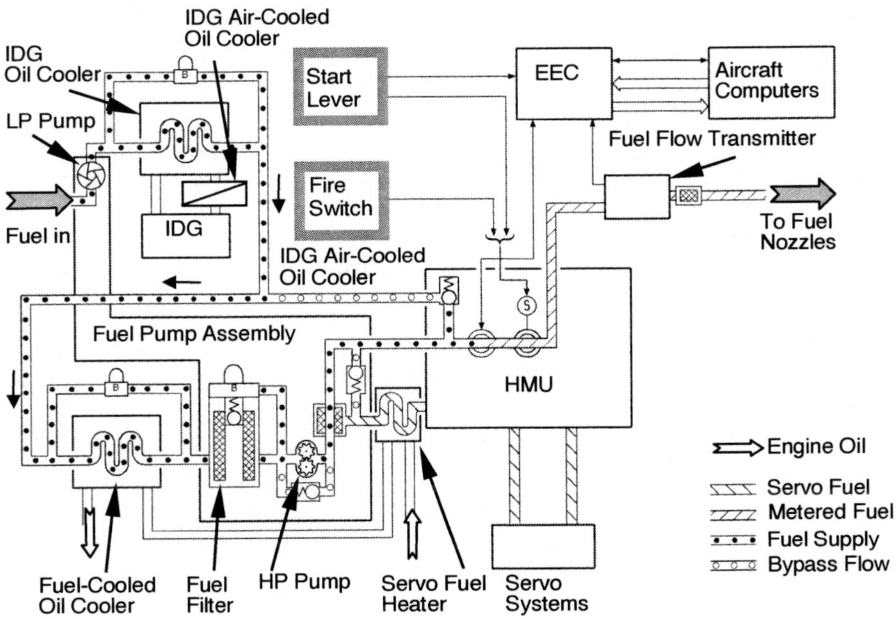

Fig. 4.1 Fuel distribution system of a CFM56-7B (© Boeing)

engine and within the fuel metering device. In the system of the CFM56-7B this fuel flow is called servo fuel and is extracted from the main flow within the fuel pump. From here the servo fuel flows towards the fuel metering device. To prevent a temperature drop below the freezing point of water within the system components passed by the servo fuel, it is heated in an additional servo fuel heater before it enters the fuel metering device. The servo fuel heater is a special feature on CFM56 engines and some CF6 versions. It is heated by engine oil.

After passing the high pressure pump, the main fuel flow enters the fuel metering device. The name used for this component depends on the engine manufacturer and the functions of this unit. GEAE and CFM use the name hydromechanical unit (HMU). Pratt & Whitney and Rolls-Royce (on some Trent versions) use the name fuel metering unit (FMU). Within the HMU or FMU the fuel metering is done with the appropriate valve. The shut-off valve for the fuel flow is also located in the fuel metering device.

From the fuel metering device the fuel supply line is routed towards the fuel manifold on the combustion case. In the supply line the fuel flow transmitter is located downstream of the fuel metering device. The wiring of the fuel flow transmitter is usually connected to the FADEC computer.

On the combustion case the fuel manifold is routed around the case to supply all fuel nozzles. Through the 20 or more fuel nozzles (the number depends on the size of the engine) the fuel is injected into the annular combustion chamber.

4.2.2 Operation

As stated above, the two-stage pump with a gear type HP stage is the most commonly used fuel pump type. The centrifugal LP stage ensures a pressure level at the HP stage inlet which prevents the occurrence of cavitation in the HP stage under all possible operating conditions. Cavitation would cause damage to the pump gears and therefore it would shorten the lifetime of the fuel pump. The worst case would be a failure of the boost pumps of the aircraft fuel system. Under this condition the LP stage is able to provide the supply of the HP stage with an adequate pressure.

The high pressure stage delivers the fuel to the metering device with the pressure necessary to inject the fuel into the combustion chamber. The fuel pressure changes with the N2 speed and reaches values up to 69 bar (1000 psi).

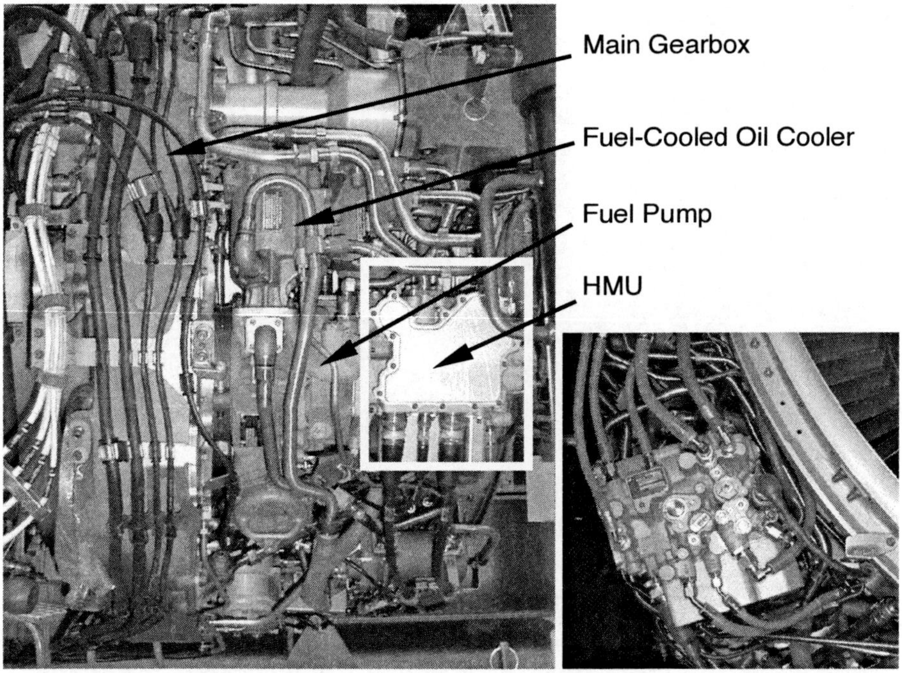

Fig. 4.2 Fuel pump and HMU installed on a CFM56-7B. The picture on the right shows a rear view of the HMU with the servo fuel lines for the various actuators on the engine (© LTT)

The fuel filter often has a common housing with the low pressure stage and the high pressure stage of the fuel pump. This design is used on the CFM56 engines and in principle on CF6 engines (with the filter bolted to the pump). On V2500 engines and on Rolls-Royce engines the fuel filter is located in the housing of the separate mounted fuel-cooled oil cooler. A design feature, which leads on the CFM56 engines to a very compact system design, is the mounting location of the heat exchangers on the fuel pump. Figure 4.2 shows this installation on a CFM56-7B.

On a typical turbofan engine, the fuel flow is high enough to cool the engine oil with the cold fuel. Therefore the oil coolers on turbofan engines are usually fuel-cooled oil coolers. On the other hand there is a requirement to heat the fuel after a certain flight time, because the fuel temperature in the wing decreases steadily during the flight and at temperatures below 0°C the free water in the fuel freezes. These ice particles would block the fuel filter. This is the reason why the fuel-cooled oil cooler is located upstream of the fuel filter.

On many engine types the fuel metering device is installed on the fuel pump. But it may also be installed on the gearbox near the fuel pump.

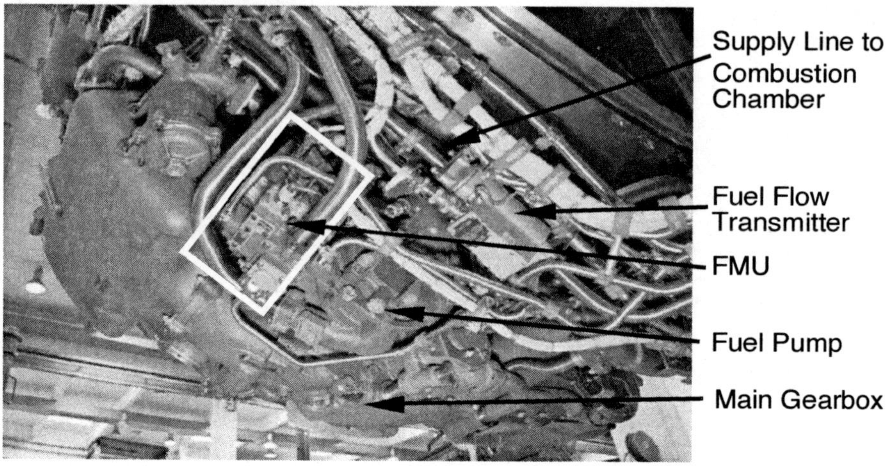

Fig. 4.3 An FMU installed on a V2500-5A (© LTT)

The simplest fuel metering device on a FADEC-controlled engine is the FMU as used on the V2500, some Rolls-Royce Trent and the Pratt & Whitney engines. Figure 4.3 shows such a FMU installed on the fuel pump of a V2500. Its main components are the fuel metering valve with bypass valve and the high pressure fuel shut-off valve. The fuel metering valve is controlled electrically by the EEC via a dedicated servo valve within the FMU. The high pressure fuel shut-off valve is controlled electrically via the engine master switch or start lever in the cockpit. The FMU of the V2500 has a third valve, the overspeed valve, which restricts the fuel flow if an overspeed condition is reached and the EEC is unable to reduce the fuel flow with the fuel metering valve. The overspeed valve is then triggered by the EEC. The main function of the FMU is fuel metering only.

The HMUs installed on engines of GEAE and CFM have more functions than a typical FMU. The operation of the fuel metering sections is practically identical. An HMU contains the fuel metering section with fuel metering valve and high pressure shut-off valve as well as the electrohydraulic servo valves for the control of the fuel pressures to the actuators installed outside the HMU on the engine. Thus all hydraulic pressure lines for these actuators are routed from the HMU to the actuators. Figure 4.2 shows these lines connected to the HMU. The HMUs on these engines also have a mechanical overspeed governor, which acts on the bypass valve to reduce the fuel flow in case of an overspeed condition. It is a back up for the overspeed protection of the EEC. For this governor the HMU has a driveshaft connected to the fuel pump. The overspeed governor becomes active only if the EEC is unable to reduce the fuel flow with the fuel metering valve.

Figure 4.4 shows the simplified schematic of such an HMU with its main components. The schematic shows how the fuel flows through this unit and which signals are necessary for the control of the fuel metering valve and the high pressure shut-off valve. The metering valve is controlled by the EEC with a closed loop control system. The EEC controls the servo valve of the fuel metering valve electrically. It is installed in the servo valve section together with all the other servo valves. The hydraulic fuel pressure from the servo valve creates the force to move the metering valve. From the position resolver on the fuel metering valve the EEC receives the feedback signal for the valve position. For more details about servo valves see Appendix B.

At each N2 speed the fuel pump delivers more fuel than the engine can consume at this speed. The fuel metering valve limits the fuel flow to the combustion chamber with the respective cross section of its opening position. The fuel that cannot pass the fuel metering valve leaves the main flow through the bypass valve as bypass fuel back to the LP pump inlet. Before the combustion fuel leaves the HMU, it passes the open HP shut-off valve. This valve is used for the shutdown of the engine. It is controlled by the pilot with the start lever or engine master switch in the cockpit. For the closure of the HP shut-off valve, its solenoid valve is energized by putting the engine master switch to off. The servo fuel pressure from the solenoid valve closes the HP shut-off valve. During engine operation the HP shut-off valve stays open while the solenoid valve is deenergized. Through this direct control of the HP shut-off valve the pilot can override the FADEC system at any time for the shutdown of the engine.

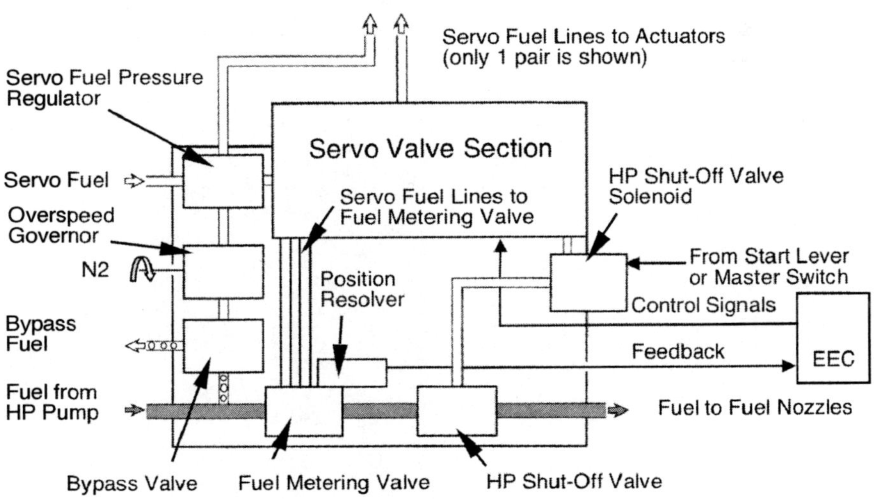

Fig. 4.4 Simplified schematic of a HMU for CFM56 engines

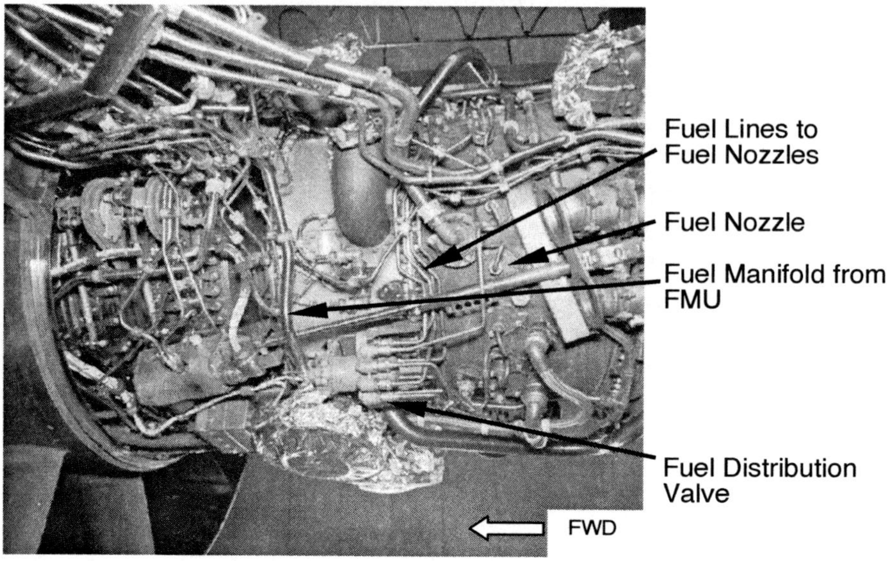

Fig. 4.5 A fuel distribution valve on a V2500 (bottom view of engine) (© LTT)

The fuel supply line is routed from the HMU or FMU along the core engine to the combustion case. The simplest design of the fuel manifold around the combustion case is used if duplex fuel spray nozzles are installed. Here each nozzle often has only one connector for the fuel tube and the manifold around the combustion case is made of two half rings.

If airspray nozzles are used, on some designs a fuel distribution valve is installed in the supply line next to the combustion case. From here several fuel supply lines connect the valve to the fuel nozzles. This has not only the function of a distributor, it is also a shut-off valve at the rear end of the

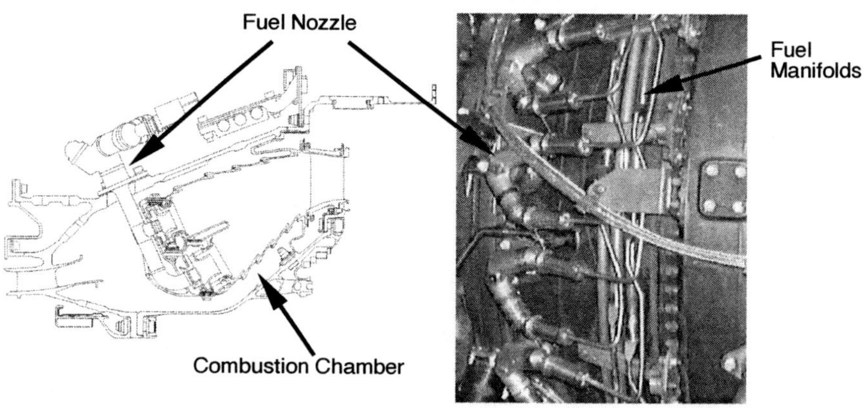

Fig. 4.6 The DAC system of CFM56 engines with 3 manifold tubes (© LTT)

fuel supply line. During engine shutdown a spring loaded valve piston within the fuel distribution valve closes. This keeps the fuel in the main fuel supply line and prevents the dripping of fuel into the combustion chamber after engine shutdown.
The third design variant uses three manifold tubes around the combustion case. It is installed on the dual annular combustor (DAC) versions of CFM56-5B and -7B engines. Figure 4.6 shows this manifold on a CFM56-7B. The combustion chamber used here with its double dome design produces less NOx emissions than the classic CFM56 combustion chamber. The DAC combustion chamber operates with a staged combustion.

Each fuel nozzle of the dual annular combustor has two duplex spray nozzles, the outer pilot nozzle and the inner main nozzle. *Staged* means that the fuel supply to the main nozzle can be switched off with a valve installed at the inlets of the three manifold tubes during intermediate power settings and idle. The main nozzles are supplied in two groups by two manifold tubes. In specific operating modes half of the main nozzles or all of them are switched off by the EEC. This operating principle creates smaller flame zones. Thus such a combustion chamber produces less NOx compared to a classical one during cruise and intermediate power settings.

The fuel nozzles of the TAPS combustor (Twin Annular Premixing Swirler) installed in the GEnx have only one fuel manifold connection. It is a single annular combustor design. The fuel is sprayed into the combustor through two discharge openings in the nozzles. Valves integrated into the fuel nozzles control the fuel flow through these passages. These valves are controlled with fuel pressure by the fuel nozzle controller. For this control pressure from the fuel nozzle controller two control lines are connected to each fuel nozzle. Thus one fuel manifold for the fuel supply of the nozzles and two signal lines are installed around the combustion case.

4.3 Thermal Management for Fuel and Oil

4.3.1 Oil Coolers and Fuel Temperature

The fuel-cooled oil coolers on turbofan engines are designed with a heat transfer capacity which keeps the oil temperature and the fuel temperature within their limits. Due to the changes in fuel flow and the generated heat within the engine during operation, the oil temperature of an engine does not remain constant. But the oil cooler is able to keep the oil temperature within the set limits despite the different fuel flows and oil flows during different operating conditions of the engine.

On many engine types the oil of the integrated drive generator (IDG) is also fuel-cooled. So the heat from the IDG oil is transferred to the fuel. If this fuel is also used for the cooling of the engine oil, the cooling effect for the engine oil is reduced by the heat from the IDG. At the higher power settings this leads only to a slightly higher oil temperature. But at a power setting slightly above idle or at idle the fuel flow is very low and the amount of heat from the IDG is not reduced. It depends mainly on the electrical load on the IDG. In this operating range of the engine the fuel temperature and the engine oil temperature can exceed their limits.

The system designers of the various engines in operation have developed different system features to cope with this problem. We will look at 4 different fuel systems as examples. These systems belong to the following engines:

- CFM56-7B
- CFM56-5A/-5B/-5C
- PW4000
- V2500

4.3.2 The Oil Cooling System of the CFM56-7B

A very simple design is used on the CFM56-7B. Here an air-cooled IDG oil cooler is installed into the IDG oil system as the first cooler to be passed by the IDG oil. In Fig. 4.1 the position of this cooler in the fuel system is shown. The second cooler to be passed at a then lower temperature level is the fuel-cooled IDG oil cooler. This cooler arrangement keeps the amount of heat transferred to the fuel low enough to keep engine oil and fuel temperatures within the limits during all operating conditions. No control devices are used in this system.

4.3.3 The Oil Cooling System of the CFM56-5A, -5B and -5C

In the fuel systems of the CFM56-engines of the -5A, -5B and -5C series the bypass fuel of the HMU is used for the cooling of the IDG oil. Figure 4.8 shows the location of the IDG oil cooler in the fuel system. From the IDG oil cooler this fuel flows into the main low pressure fuel flow upstream of the fuel-cooled oil cooler. Thus the temperature of the bypass fuel downstream of the IDG oil cooler influences the cooling effect of the fuel-cooled oil cooler for the engine oil. The mixing of this hot fuel with the cold main flow from the tank basically prevents excessive fuel temperatures within the engine fuel system.

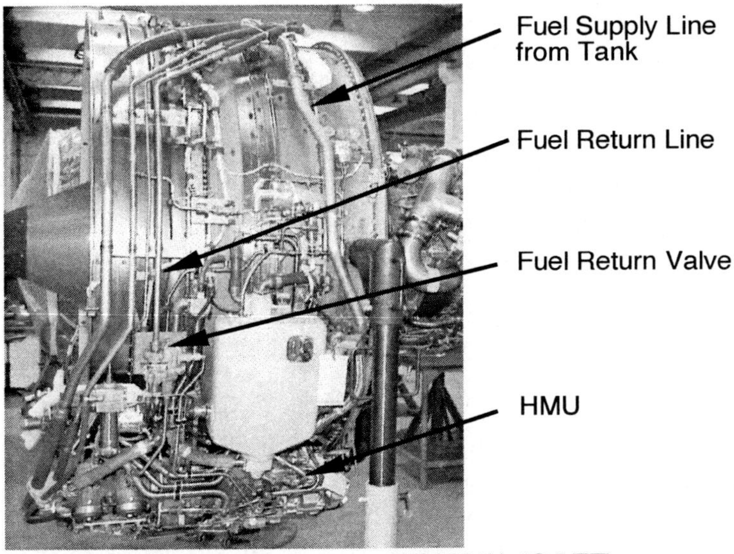

Fig. 4.7 The fuel return valve on a CFM56-5A (© LTT)

But the engine oil temperature can reach high values if the fuel flow is low and the heat from the IDG oil is on a high level. To cope with this condition, these engines have a fuel return valve installed in the fuel

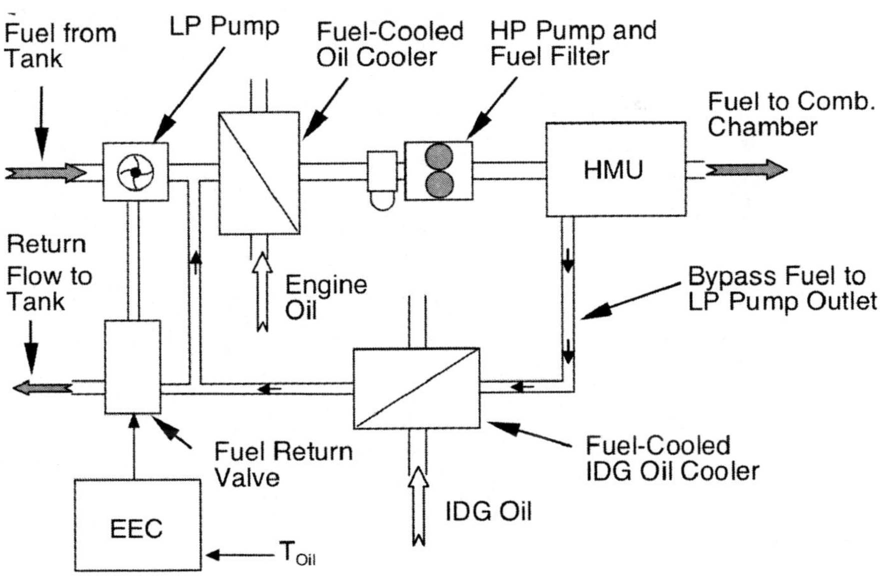

Fig. 4.8 Cooling of the IDG oil in the fuel system of a CFM56-5A,-5B and -5C

line downstream of the IDG oil cooler (see Fig. 4.7). If the engine oil temperature is too high, the EEC opens the fuel return valve and a defined percentage of the HMU bypass flow is routed directly into the aircraft fuel tank. At the inlet of the fuel-cooled oil cooler this amount of returned fuel is substituted by cold fuel from the fuel pump, which improves the cooling effect of this cooler. The fuel return valve is controlled by the EEC using a very simple schedule based on an oil temperature threshold only. The fuel temperature is not monitored in this fuel system .

4.3.4 The Oil Cooling System of the PW4000

The oil system of the PW4000 has an air-cooled oil cooler additionally to the fuel-cooled oil cooler. In this design, as shown in the schematic in Fig. 4.9, the fuel-cooled oil cooler is the primary oil cooler. It cools the engine oil and also the IDG oil. The air-cooled oil cooler is normally not put into operation because its cooling airflow is shut off by an air control valve. In the IDG oil circuit an air-cooled oil cooler is installed, too. It improves the cooling of the IDG oil during engine low power operation and when excessive IDG oil temperatures occur.
The purpose of this engine oil cooler arrangement is to prevent excessive heating of the fuel and to keep the engine oil temperature within the limits. Therefore the EEC monitors the engine oil temperature and the fuel

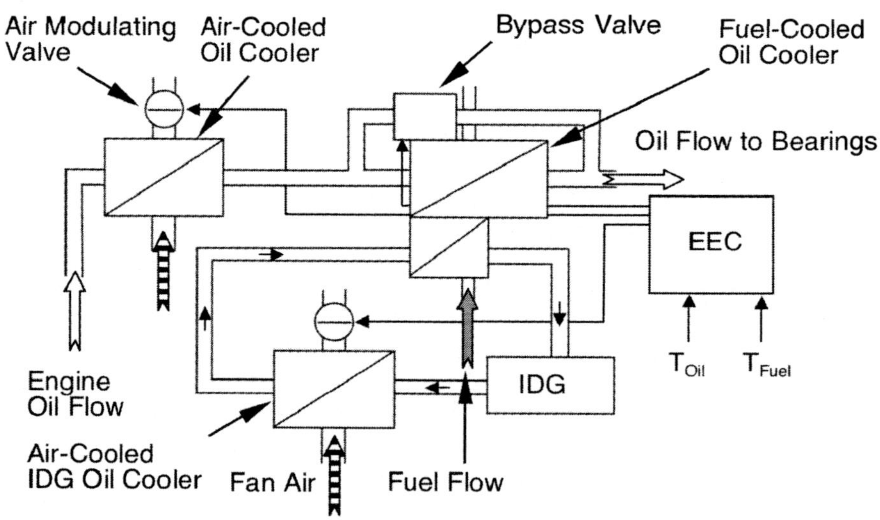

Fig. 4.9 Oil cooler arrangement of a PW4000

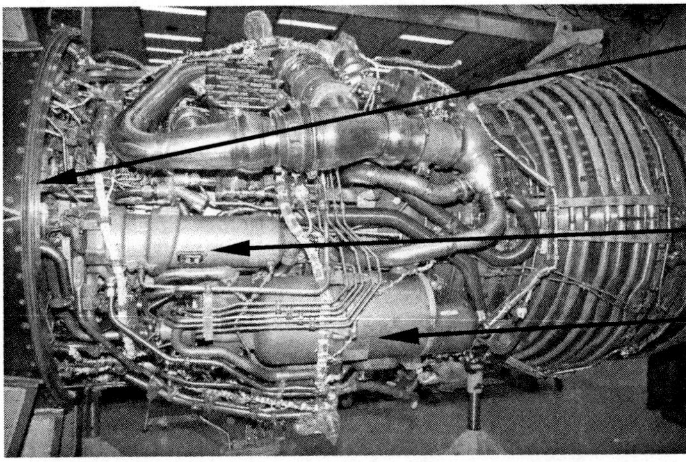

Fig. 4.10 The fuel-cooled oil cooler of a PW4000 (© LTT)

temperature. If the fuel temperature in the engine system reaches the threshold temperature, the EEC can reduce the amount of heat transferred into the fuel. For this purpose it switches the oil flow bypass valve of the fuel-cooled oil cooler into the bypass position and opens the air control-valve of the air-cooled oil cooler. With the operating air-cooled oil cooler the cooling of the oil is ensured and with the bypass valve in the bypass position, a further heating of the fuel by the engine oil is prevented.

In case of a high engine oil temperature the EEC opens the air control valve of the IDG air-cooled oil cooler. This reduces the temperature of the IDG oil before it reaches the fuel-cooled oil cooler. So the cooling of the engine oil in the fuel-cooled oil cooler is improved.

4.3.5 The Oil Cooling System of the V2500-A5

The V2500 is also an engine with a fuel-cooled IDG oil cooler. In the oil system of this engine an air-cooled oil cooler is installed as an additional cooler to the fuel-cooled oil cooler. The fuel system has the fuel return feature like the CFM-engines but the valve used on the V2500 has some additional functions for the change of flow directions of the relevant fuel flows. It is called the fuel diverter and return valve (FDRV).

The cooling fuel for the IDG oil is normally cold fuel from the LP pump. The EEC monitors the engine oil temperature, the fuel temperature and the IDG oil temperature. It can operate the FDRV to send the IDG cooling fuel back to the aircraft tank or to reroute the flow of the IDG cooling fuel and the FMU return fuel within the fuel system. To make this

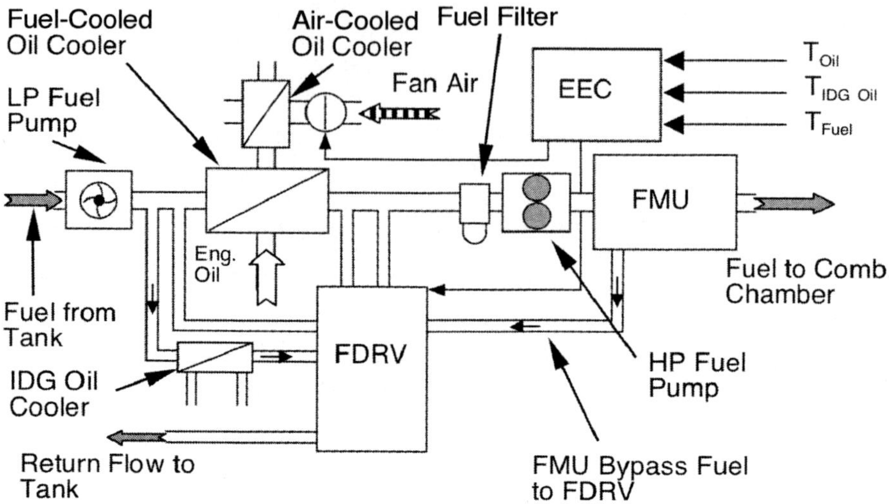

Fig. 4.11 Engine oil and IDG cooling system of a V2500

possible, the FDRV has five connections to the engine fuel system. A simplified system schematic is shown in Fig. 4.11. To prevent excessive oil or fuel temperatures, the EEC can open the air valve of the air-cooled oil cooler for the precooling of the engine oil.

To switch the FDRV four operating modes are programmed in the EEC. Two are with fuel return to the aircraft and two are without fuel return.

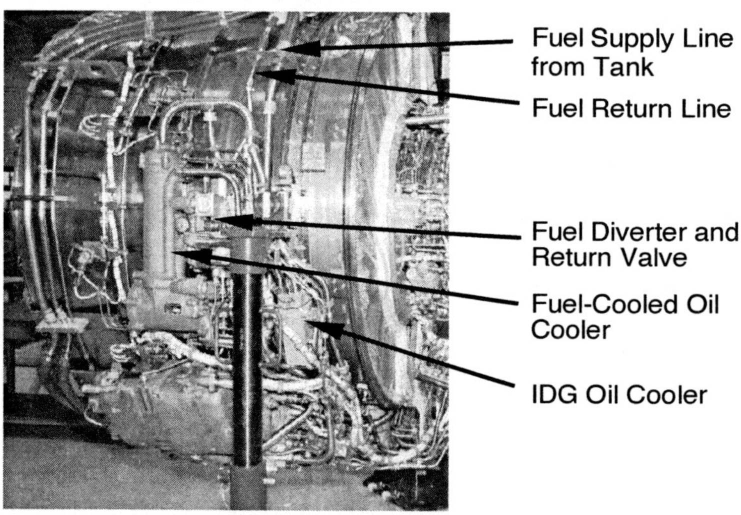

Fig. 4.12 The fuel diverter and return valve on a V2500 (© LTT)

Fig. 4.13 The air-cooled oil cooler of a V2500 (© LTT)

One of the modes with fuel return is the normal mode without any cooling by the air-cooled oil cooler. In the other modes the operation of the air-cooled oil cooler is possible. The EEC modulates the airflow through the air-cooled oil cooler to keep the temperatures within the limits and to minimize the cooling airflow. Because the cooling airflow is an air loss to the fan airflow, the amount of cooling air is minimized to keep engine efficiency as high as possible.

5 Engine and Fuel Control System

The following description of the engine and fuel control system explains the principles of fuel and engine control as well as the architecture of the most commonly used designs of fuel control systems. All modern turbofan engines are controlled by full authority digital engine control (FADEC) systems. Despite this fact, large numbers of older hydromechanically controlled engines are still in use. To give an overview of these kind of systems, they are described as well.

5.1 Main Tasks of the System

The main task of the engine and fuel control system is the metering of the fuel flow to the combustion chamber under the operating conditions of steady state, acceleration and deceleration. The system must achieve this with the aim to keep the engine on the thrust level demanded by the pilot. To prevent the exceedance of the operating limits, limiting functions ensure the operation within the limits for shaft speeds, temperature and pressures. To accomplish this task the engine control system also controls the functions of the other engine systems. These systems are known as the engine subsystems for the description of the fuel and control system.

5.2 Speed and Thrust Control

For the control principle of a turbine engine the shaft speed and the thrust are the two most important parameters. The simplest control system is a shaft speed control system. As long as the power lever angle is constant, it keeps the shaft speed constant. A typical application for a constant shaft speed control system is the APU. If an aircraft engine is controlled by a shaft speed control system, the thrust will change with changing air density and the pilot has to readjust the thrust levers to keep the required thrust constant. For the operation of an aircraft engine a thrust control system is most suitable. Here the pilot can directly select the thrust needed for a flight maneuver and the control system keeps this thrust constant.

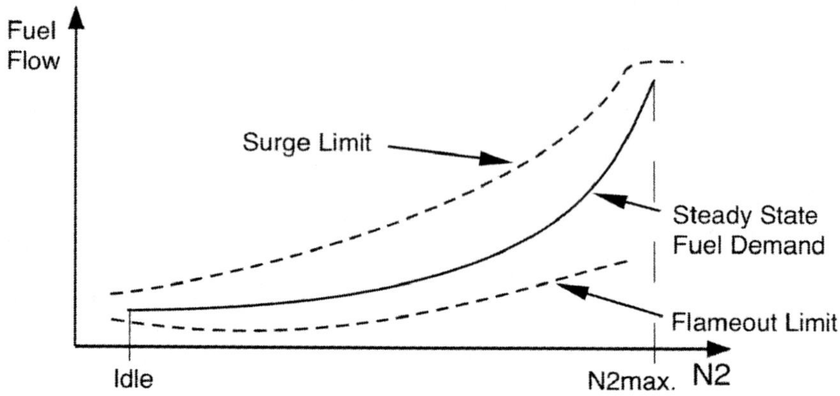

Fig. 5.1 Fuel demand curve of a turbine engine

5.2.1 Shaft Speed Control

In a speed control system the shaft (or rotor) speed is the main control parameter. The fuel control unit (FCU) uses a speed governor to adjust the shaft speed to the value demanded by the input of the power lever. For the calculation of the fuel flow the FCU follows the fuel demand function of the specific engine. In Fig. 5.1 this function is shown based on the N2 speed of a twin-spool engine. It contains the graph for the steady state fuel demand at the N2 speed and the two limit lines, one for the surge limit and the other for the flameout limit. The fuel flow for the acceleration must be between the steady state line and the surge limit. For the deceleration the fuel flow must be between the steady state line and the flameout limit.

To adjust the calculated fuel flow to the current air density at the combustor inlet, the FCU needs at least the inputs of the compressor inlet temperature (CIT) and the compressor discharge pressure (CDP or Ps3) additionally to the N2 speed and the power lever input. With these inputs a simple speed control system can operate. Figure 5.2 shows these inputs into an FCU for a speed control system.

To operate as a control unit for a turbine engine the FCU must have a minimum of functions. These are established in functional subsections of the FCU. As an example let's have a closer look at the hydromechanical FCU. For the operation as a speed control device it needs the following subsections:

- The fuel metering section
- The governing section
- The limiting section

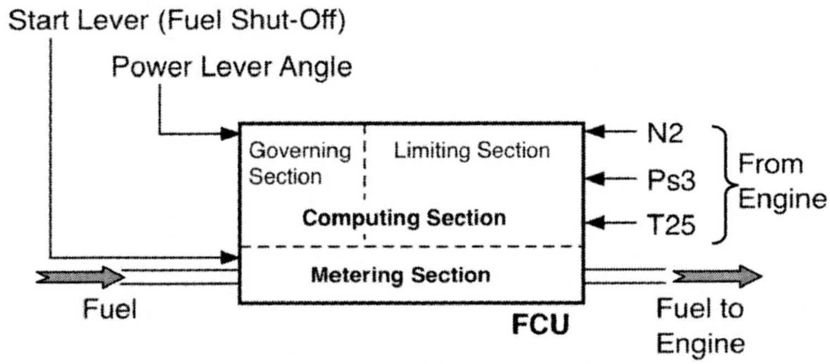

Fig. 5.2 Functional sections of a fuel control unit for a speed control system

The governing section and the limiting section together form the computing section of the control unit. Figure 5.2 shows the FCU sections schematically. The design of the metering section is the same as shown in the previous chapter on the HMU. The fuel metering valve is moved by control pressures from the computing section. For the governing of the shaft speed, the governing section contains a flyweight governor, which receives the speed demand from the power lever. It controls an opening or closing of the fuel metering valve to hold or reach the demanded speed. The control signal from the governor is modified by the limiting section to limit the opening or closing rate of the fuel metering valve.

5.2.2 Thrust Control

For thrust control the control system uses a main control parameter, which is directly related to the thrust of the engine. This parameter can be the N1 speed or the engine pressure ratio EPR. Which of the two is used depends on the philosophy of the engine maker. GEAE and CFM use the N1 speed; Rolls-Royce and Pratt & Whitney use the engine pressure ratio for this purpose.

If the N1 is the main control parameter, the control system keeps the airflow through the engine constant by adapting the N1 speed to the changing density of the ambient air. For this purpose it must receive the information about the air density from the fan inlet.

Such a control system can be based on the speed control system for the gas generator, which is supplemented by the density correction for the fan inlet parameter. The inputs necessary for this density correction are the pressure and temperature from the fan inlet. But this system can only

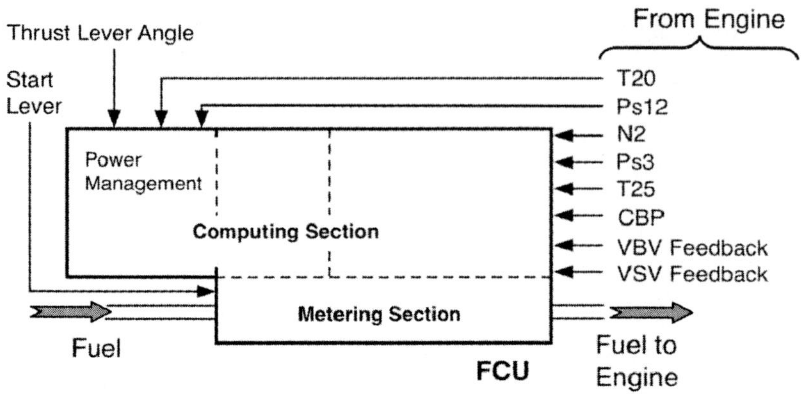

Fig. 5.3 Inputs for a FCU with thrust control capability for a twin-spool engine

adjust the HP rotor speed to the changing air density. Figure 5.3 shows the input signals of the CFM56-3 hydromechanical control as an example. This unit is an N2 control unit, because the N2 is the only shaft speed information for this control unit. The matching between the two shaft speeds changes with the deterioration of the engine, and the actual N1 speed will always be a few percent below or above the correct N1 for a thrust setting if the control unit controls the N2 only. For the precise control of the airflow (and the thrust), it is necessary to control the fan rotor speed N1 precisely at the correct value for the demanded thrust. Control systems with hydromechanical FCUs use electronic supervisory control units as an N1 controller additionally to the N2-controlling FCU to solve this problem. Such a supervisory control unit for example is the Power Management Control of the CF6-80C2 (PMC version) or of the CFM56-3.

If the control system uses the EPR as the main control parameter, it keeps the thrust constant by holding the EPR constant. Thus it also keeps the airflow of the engine constant. The N2-controlling FCU is overridden by an EPR controlling electronic supervisory control unit. This EPR control principle is used on versions of the JT9D-7R4 and the RB211-524.

With such a system a fixed relation between the thrust lever angle (TLA) and the demanded thrust exists as shown in Fig. 5.4. For a specific thrust the pilot always selects the same TLA. Because the control system can't control the thrust directly, the supervisory control unit calculates the value of the main control parameter (N1 or EPR) in relation to the demanded thrust. This is done in consideration of the actual atmospheric conditions. The result of this calculation is called N1command or EPR-command. For this calculation function the term *power management* is

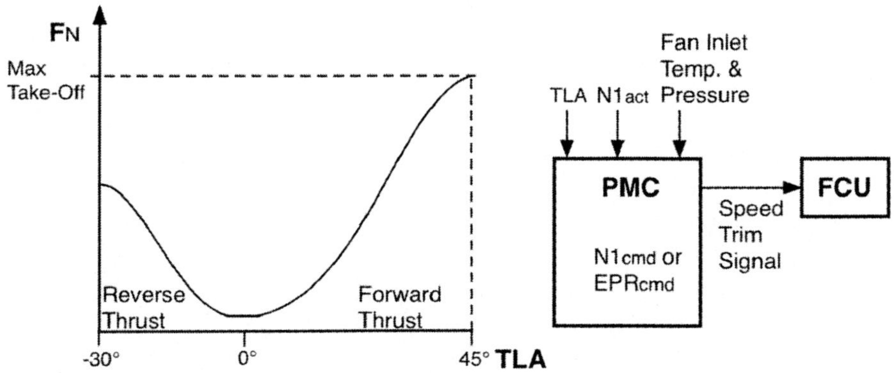

Fig. 5.4 The relation between the thrust lever angle and the demanded thrust is fixed through all flight phases. The PMC calculates the related N1cmd or EPRcmd

used. On engines of GEAE and CFM the supervisory control units are named following this term. Here this unit is called *power management control* (PMC). The PMC compares the actual N1 to the N1-command. If there is a difference, the PMC overrides the speed control of the FCU. To do this it changes the N2 speed until the actual N1 speed equals the N1 command. In an EPR related system the supervisory control unit operates similarly but compares the actual EPR to the EPRcommand.

These examples of thrust control systems were the last designs using hydromechanical control systems. Engines developed later use pure electronic control systems. Because the digital computer is the only control unit and it has full authority over engine control, these systems are called *Full Authority Digital Engine Control systems*. The abbreviation of this name, FADEC, is a well-known term in aviation. The main steps in the control system evolution are shown in Fig. 5.5.

FADEC systems are used as N1 control systems or EPR control systems for constant thrust control. The main advantage of the use of FADEC systems for engine control is the ability of the systems to process more inputs than a hydromechanical system. One benefit of this attribute is the behavior of an FADEC-controlled engine during accelerations. The system keeps the dN2/dt constant over the whole engine life regardless of deterioration. Another consequence is an actuator positioning and fuel flow control which is more precisely adapted to the actual engine operating condition. The huge amount of data available in the computer is also used for performance and engine health monitoring, the so-called *engine condition monitoring* (ECM).

Due to the hardware and software design an FADEC system has a higher level of redundancy compared to the hydromechanical system.

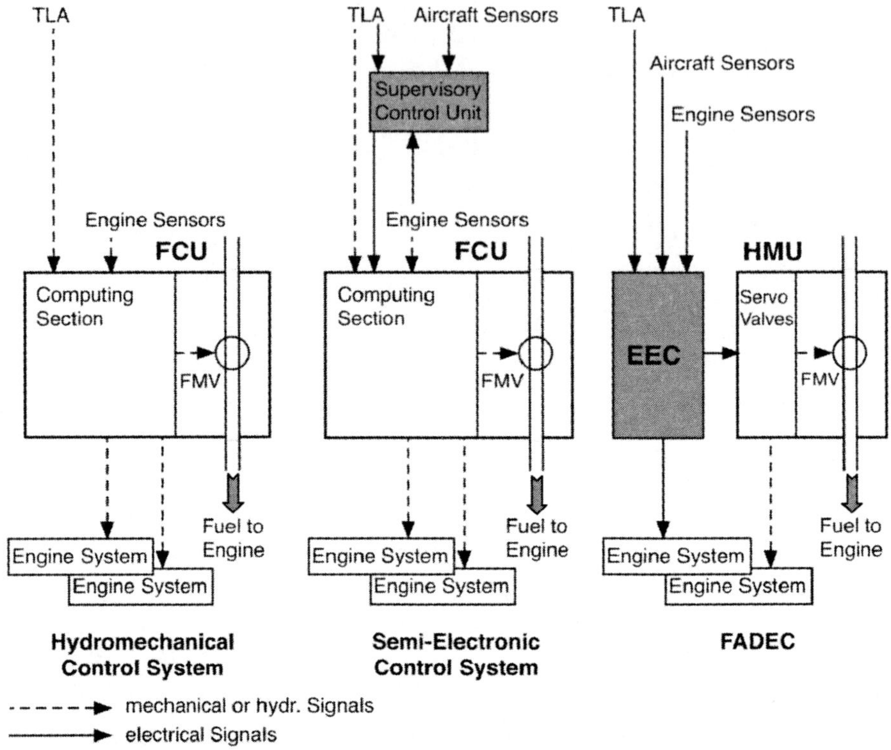

Fig. 5.5 Evolution of the engine control systems from the hydromechanical system to the FADEC system

The whole FADEC system is designed to be fault tolerant. As a result of this design principle several data inputs can be lost or components can fail without an effect on the engine operation.

5.2.3 Engine Thrust Ratings

The thrust an engine can generate has a range from idle to the maximum certified thrust. Within this range the main thrust levels for flight operation are designated. These designated thrust levels of an engine are called *thrust ratings*. The thrust level selected by the pilot, also between the designated ratings, is called *thrust setting*.

5.2.3.1 Rating Structure of a Commercial Turbofan Engine

The highest thrust ratings on the rating scale are the two certified operating limits of the engine. These are *maximum take-off thrust* and *maximum continuous thrust*. On engines certified according to EASA CS-E or FAR33 the use of maximum take-off thrust is limited to a duration of 5 minutes per flight cycle. The thrust levels of the ratings below maximum continuous thrust are determined in accordance with the requirements of the aircraft. These ratings are:

- Maximum climb thrust
- Maximum cruise thrust
- Maximum reverse thrust

The maximum reverse rating is mainly an N1 speed limit. In Fig. 5.6 the ratings are shown on a scale with idle ratings at the lower end of the scale. Several different idle ratings are used on turbofan engines. They are no thrust ratings because the idle ratings are shaft speed related. In the idle range the control system uses the HP shaft speed as the control parameter.

One rating all engines have in common is the minimum idle rating. At this rating the HP shaft runs at its lowest possible speed to minimize the generated thrust. This shaft speed is slightly higher than the minimum speed from which the engine can accelerate. The idle speed in general is not a constant value. It increases with decreasing air density. The minimum idle speed on FADEC-controlled engines is often a fixed value over a broad OAT range.

For the acceleration from minimum idle speed to 95% of maximum take-off thrust a time limit is established in the certification standards. This limit is necessary to ensure the immediate availability of the thrust during a balked landing maneuver. The acceleration time most turbofan engines need from minimum idle to 95% of maximum take-off thrust is too long to comply with certification standards. To fulfill this requirement, a higher idle speed is used during flight, the so-called flight idle or high idle speed.

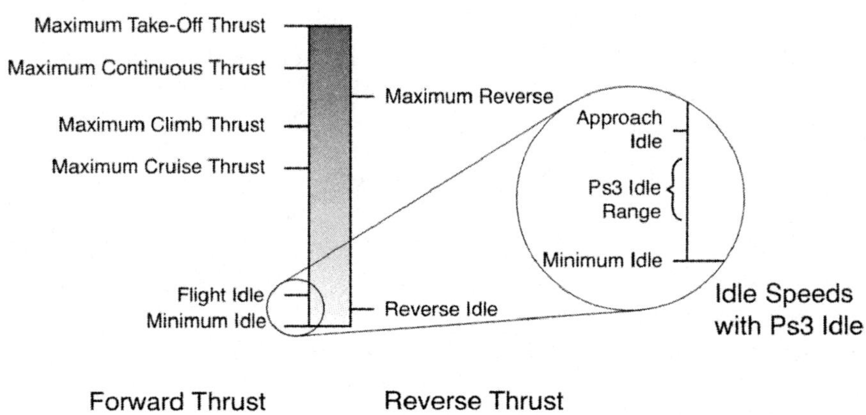

Fig. 5.6 Thrust ratings and idle speeds

On ground the minimum idle speed is used. This ensures that the aircraft can be taxied without an excessive use of brakes. In the simplest system designs the switching to flight idle is done by the air/ground logic of the aircraft. Thus this higher idle speed is valid during the whole flight (idle control system of the B737-300/-400/-500).

The higher flight idle thrust is a disadvantage during the descent of the aircraft in clean configuration. To solve this problem, more sophisticated systems also use the flap position as a signal for switching to flight idle. With such a system the flight idle is only used in the approach configuration. It is then called approach idle.

Due to the fact that a FADEC system can process more inputs, such systems use some additional idle ratings. Their use depends on the condition of the related airframe systems.

One of these ratings is Ps3 idle (but not limited to FADEC engines). At minimum idle during flight, the supply pressure for the aircraft pneumatic system can be too low. To keep the air pressure for the supply of the environmental control system high enough, the Ps3 must be held on a sufficient level. This is ensured by the use of Ps3 as the control parameter during idle setting while the pneumatic system is supplied by the engine. The EEC calculates the required Ps3 from the bleed air demand of the aircraft pneumatic system. The N2 speeds reached during Ps3 idle are lower than the approach idle values.

The other additional idle rating used is the one influenced by the oil temperature as used on the CFM56-5A/-5B/-5C and the V2500. When the engine runs at minimum idle or at Ps3 idle and the oil temperature cannot be kept under the programmed limit by the cooling system, the EEC calculates an idle speed demand as a function of the oil temperature. If this idle speed demand is higher than the present idle speed, the N2 speed increases. The higher fuel flow for this higher idle speed improves the oil cooling. The N2 speed limit is at or slightly higher than approach idle.

5.2.3.2 Flat Rating

A commercial turbofan engine is operated as a flat-rated engine. This means that the maximum take-off thrust is available up to a specific OAT. This temperature is called the flat rate temperature (FRT). The thrust/OAT curve shows a flat range until it reaches the flat rate temperature (seeFig. 5.7). Up to the FRT the EGT increases with the increasing OAT at the constant thrust. In the range beyond the FRT the thrust decreases with increasing OAT, shown by the negative slope of the thrust curve. If the thrust setting in this OAT range follows the curve, the EGT remains constant despite the increasing OAT.

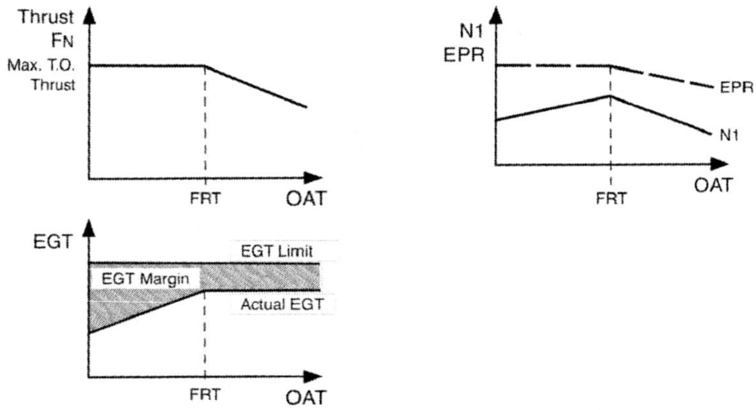

Fig. 5.7 Thrust, EGT and EPR over the OAT for a flat-rated engine. The EGT margin has its lowest value at the FRT or beyond that temperature

The advantage of this rating concept is the availability of the maximum take-off thrust over a wide OAT range combined with the low EGT values at low OATs. The latter leads to lower thermal loads of the engine. The whole operation under the flat rate concept results in longer engine lives, because the average EGT over the operating time of the engine is lower compared to a full-rated engine. As a full rated engine the engine would operate at the maximum permissible EGT during each take-off.

5.3 Hydromechanical Control Systems

This chapter gives a brief overview of the system layout of two thrust control systems as examples. Both systems have a supervisory engine control unit and the N1 is the primary control parameter. The systems belong to the CFM56-3 and the CF6-80C2.

5.3.1 Engine Control System of the CFM56-3

On the CFM56-3 the control system consists of two control units, the MEC and the PMC with their associated sensors. The hydromechanical control unit of the CFM56-3 is called Main Engine Control (MEC). This name is used by CFM and GEAE for these units because they operate as a fuel control unit and as a control unit for the subsystems as well. On the CFM56-3 these subsystems are the VBV system, the VSV system and the active clearance control for the HPT. The N2 is the only shaft

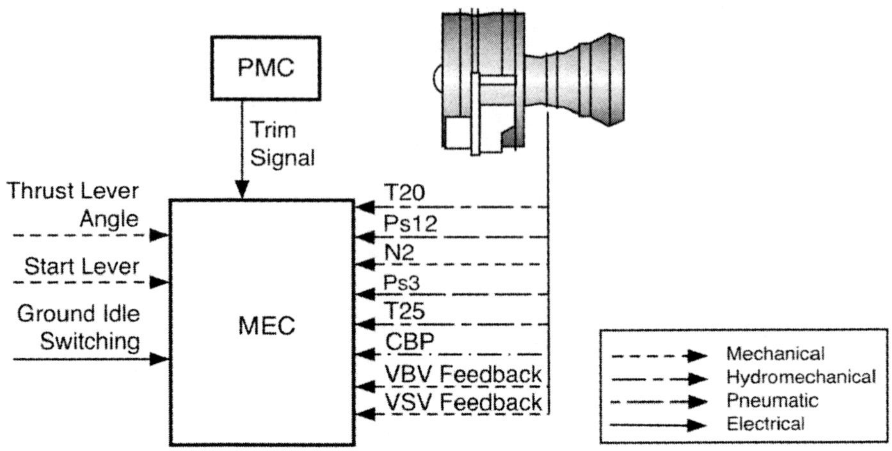

Fig. 5.8 Input signals to the MEC of a CFM56-3

speed input. An air density compensation is possible, because it receives the Ps12 and the T20 from the fan inlet. If the MEC operates without the PMC, it operates as a thrust control unit based on N2 speed. So it's possible to operate the engine without the PMC function. But the pilots must bear in mind that the engine has a slight deviation from the correct N1 at a certain TLA. Figure 5.8 shows the system schematically with the related inputs. The customer bleed pressure (CBP) is used for the detection of the operation of the bleed air supply to the airframe. With this input the MEC reschedules the acceleration fuel to keep the acceleration time nearly constant despite the operating bleed system. For the change to ground idle an

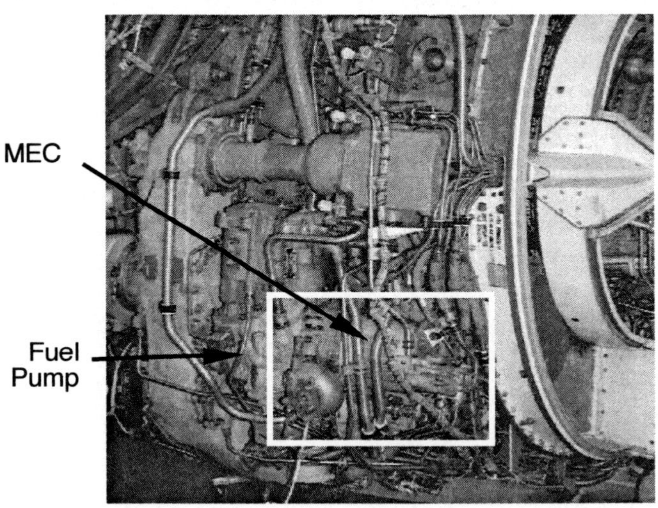

Fig. 5.9 MEC on the left side of a CFM56-3 (© LTT)

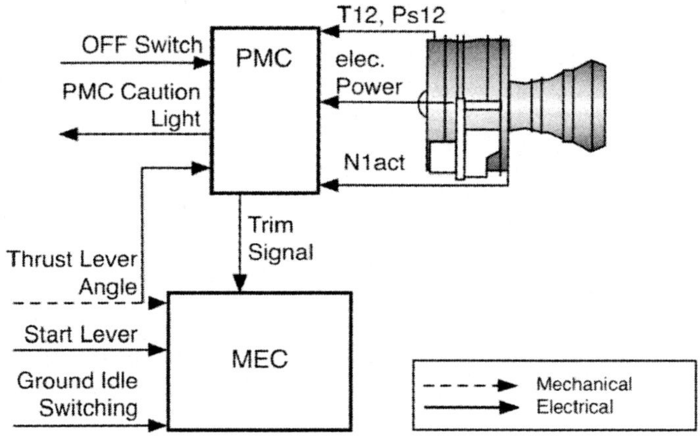

Fig. 5.10 The PMC system of a CFM56-3

electrical switching signal from the aircraft's air/ground logic is sent to the MEC. In Fig. 5.9 the location of the MEC on a CFM56-3 is shown. At the rear face of the MEC the fuel lines to the actuators of the subsystems are visible. All these actuators are operated by using fuel pressure from the MEC.

The PMC is an electronic component that needs an electric power supply. On the CFM56 the PMC is supplied by a small alternator installed on the accessory gearbox. The alternator starts the supply of the PMC at an N2 below idle. With this alternator the system is independent of an aircraft power supply. The inputs into the PMC necessary for the power management and the control function are shown in Fig. 5.10. After calculating the

Fig. 5.11 The PMC on the right side of a CFM56-3 (© LTT)

N1cmd the PMC compares the N1cmd to the actual N1. If an error exists, the PMC sends a trim signal to the MEC. This trim signal of the PMC is a current to the torque motor installed within the MEC. With this torque motor the PMC changes the speed setting on the speed governor of the MEC a few percent up or down and the MEC accelerates or decelerates the engine to reach the N1cmd. Because the MEC has controlled the engine at a certain thrust lever angle to an N2 speed at which the N1 is near the N1cmd, the PMC needs only this small trim range for its control function.

Due to its ability to control the N1 very precisely the PMC prevents the exceedance of maximum take-off thrust by advancing the thrust lever too far. This also protects the engine from excessive gas temperatures on hot days.

5.3.2 Control System of the CF6-80C2 (PMC Version)

The system architecture is very similar to that of the CFM56-3. But on the CF6-80C2 the PMC additionally uses data from the airframe systems and the MEC has an N1 input. Both units are shown in Fig. 5.13. As the static pressure at the inlet they use the ambient pressure P0 instead of Ps12 (static pressure in front of the fan). The basic functions of the MEC and the PMC are the same as on the CFM56-3 but on the CF6-80C2 these two components are more sophisticated. For the MEC this means that it has an N1 and an N2 governor. In the idle speed range the N2 governor controls the fuel flow and above idle the N1 governor controls the fuel flow.

The MEC is able to control an N1 speed as a function of the TLA, but it operates with the inputs measured by it. This also leads to a slight deviation from the correct N1 speed for the selected thrust.

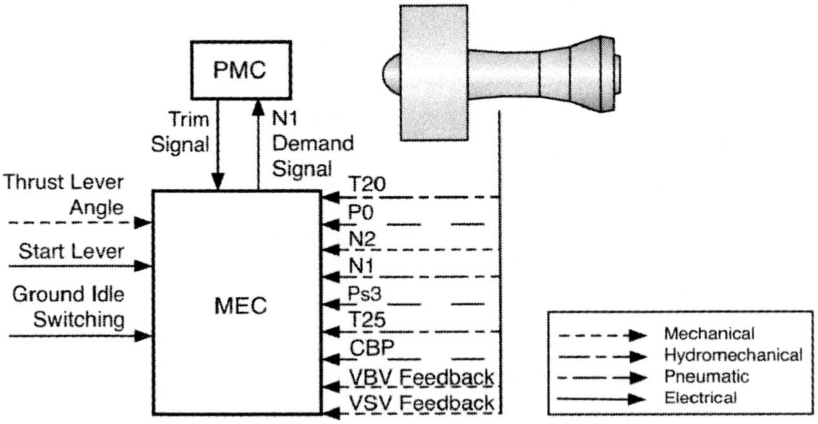

Fig. 5.12 Input signals to the MEC of a CF6-80C2

Fig. 5.13 The left picture shows the MEC of a CF6-80C2 on the gearbox below the core engine. The right picture shows the PMC on the fan case (© LTT)

The PMC uses the more precise data from the air data computer (ADC) including the Mach number. Therefore it is able to calculate the N1cmd very precisely. All the input data used by the PMC are shown in Fig. 5.14. The digital inputs are the values of Mach number, total air temperature and total pressure from the ADC and the bleed status from the respective aircraft computer. From the MEC the PMC receives the N1-demand signal from a built-in resolver. This is the N1-command of the MEC. The PMC uses this value to calculate the necessary change of the trim signal to the MEC. With the bleed status data the PMC is able to limit the take-off thrust to prevent excessive EGT values at take-offs with an operating engine bleed air system.

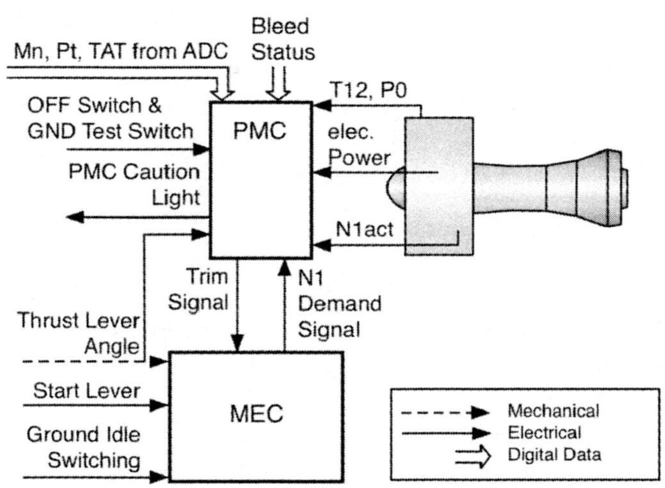

Fig. 5.14 The PMC system of a CF6-80C2. In contrast to the CFM56-3 PMC this PMC receives digital data from the airframe via data buses

5.4 FADEC System

If a full authority digital engine control (FADEC) system is used for engine control, this has the advantage that its computer can process a lot more parameters than a hydromechanical control system. Therefore this capability of the system is used for additional functions to optimize the engine operation. Additional to the basic engine control functions of power management and fuel control the following functions are usually assigned to the system:

- Monitoring and fault detection for the system and all the connected components.
- Data source for engine indication
- Data source for ECM
- Starting, shutdown and ignition control
- Control of the thrust reverser
- Automated system tests

5.4.1 System Design

The FADEC system is designed around the electronic engine control (EEC), which is the FADEC computer. The components forming the core of the system are the EEC, the sensors and the FMU or HMU. All the other components controlled by the EEC are assigned to the other engine systems like air system or ignition system. The forces that move valves or stator vanes are created by the use of fuel or air pressure. The control valves of the hydraulic part of the system can be arranged centralized in an HMU

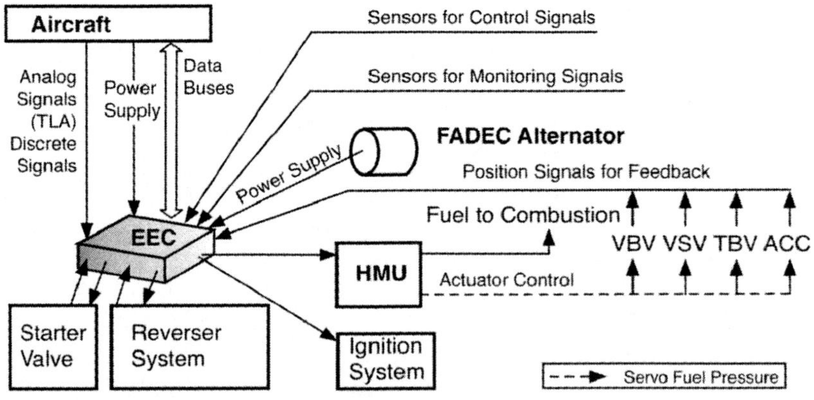

Fig. 5.15 The system architecture of a FADEC system with the centralized arrangement of servo valves in an HMU

or decentralized on the individual actuators with an FMU installed in the fuel distribution system. The centralized variant is used on engines of GEAE and CFM. Figure 5.15 shows this design principle. The decentralized variant with the servo valves located on the individual actuators is used on engines of Pratt & Whitney, IAE and some Rolls-Royce engines.

The electrical power for the system is provided by a small alternator located on the accessory gearbox as the main power supply. The alternate power supply comes from the electrical system of the aircraft. With its own alternator the FADEC system is independent of the power supply from the aircraft during normal operation. This is important if the aircraft system or the connection to it fails. This failure would have no affect on the engine operation. The alternate power supply from the aircraft is used for engine starting and in case of an FADEC alternator failure. During the start of the engine the power supply is taken over by the FADEC alternator. The amount of power the FADEC system needs is relatively low (approx. 2x300 W) because it is used only for the operation of the EEC, the sensors and the control of the servo valves and solenoid valves. In Fig. 5.16 the EEC and the FADEC alternator on a CFM56-7B are shown.

All the components of the FADEC system and the components controlled by the FADEC system are designed as line replaceable units (LRU) for quick replacement. The time needed for the replacement of such a component is 15 to 45 minutes and depends on the engine type and the component. To achieve these replacement times during real operation, the LRUs are designed for an installation without rigging or electrical adjustments.

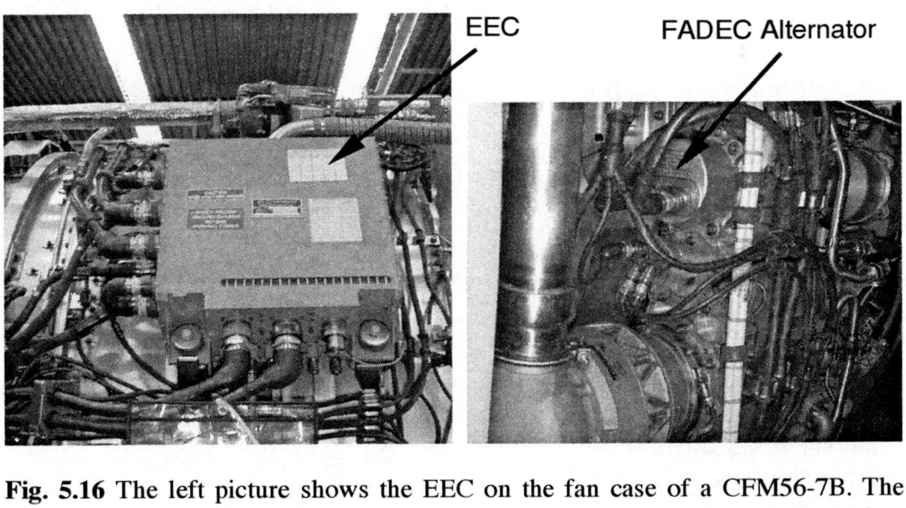

Fig. 5.16 The left picture shows the EEC on the fan case of a CFM56-7B. The right picture shows the FADEC alternator on the accessory gearbox of this engine. The two harness connectors for Channel A and Channel B are visible (© LTT)

Fig. 5.17 The EEC of the CFM56-7B. The markings show the case sections for the different EEC components (© LTT)

5.4.2 The EEC

The EEC is usually located in the fan case area of the engine. Here the ambient temperature has acceptable values during the flight for the operation of the EEC. For the redundancy of the hardware the EEC contains two identical computers called Channel A and Channel B. Each of the two channels has its own power supply with one connection to the FADEC alternator and one to the aircraft network. Figure 5.17 shows the location of the main components of a CFM56-7B EEC as an example.

5.4.2.1 Inputs and Outputs

Within the EEC the pressure transducers for the measurement of air pressures are installed. This is a similar feature of the systems used on all large turbofan engines. The pressures to be sensed are picked up by pressure probes at the respective airflow station of the engine gas path and transferred through a sensing tube to the transducer inside the EEC. To sense temperatures and the thrust lever angle (TLA) electrical sensors are used. They are installed at the sensing location.

To have redundant sensors, the temperature sensors, the TLA sensor and the pressure transducers are dual. For each sensed value a Channel A and a Channel B sensor is installed. The same principle is used for the position feedback sensors located in the actuators and valves. To have maximum hardware redundancy the sensors are connected to the EEC with separate connectors and cables. If an engine has additional monitoring sensors to

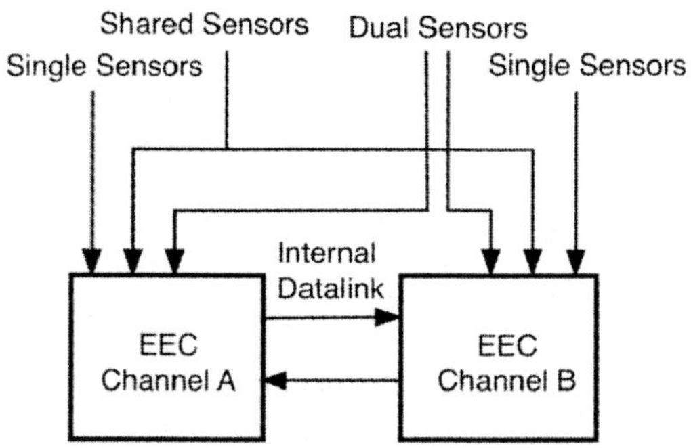

Fig. 5.18 The different sensor types connected to an EEC

sense data for ECM only with these sensors, they are single sensors and connected to one of the two channels. The fuel flow transmitter is usually a single sensor. But it is connected to both channels. That's why this installation is called a shared sensor. Figure 5.18 shows these principles of sensor inputs to an EEC. The harness connectors of the dual position sensors are shown in Fig. 5.19.

Additionally to the inputs from the various sensors the EEC receives data from the airframe systems. The air data system for example sends data like the Mach number, total air temperature and altitude. Most EECs are able to calculate these values from their own data, but the values from the air data system are more accurate.

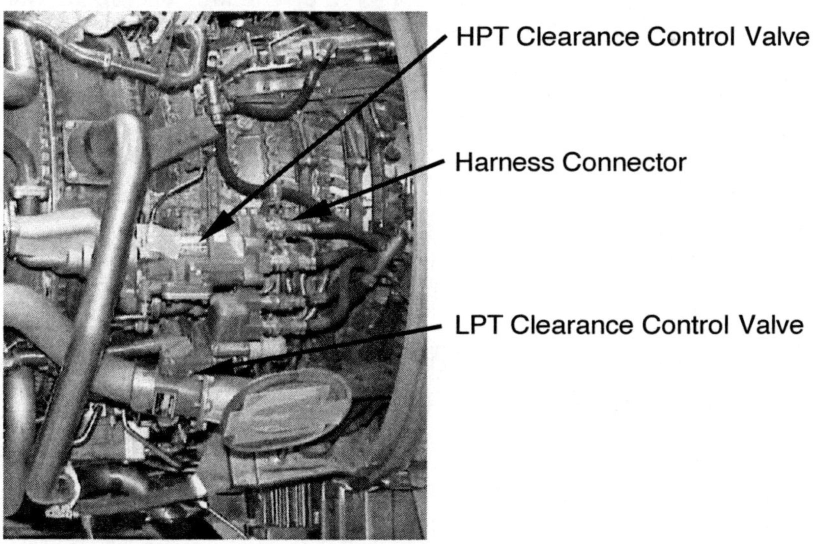

Fig. 5.19 Dual harness connections on the position sensors of the ACC valves of a CFM56-5A (© LTT)

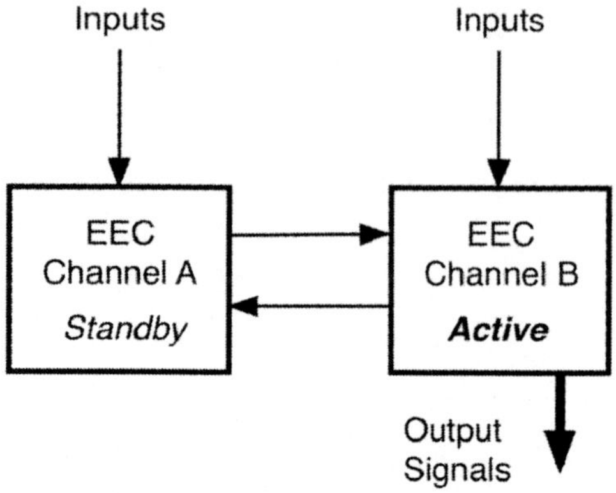

Fig. 5.20 Both channels are processing the data. Only the active channel controls the actuators with its output signals.

All outputs of the EEC are dual as well. On the output side the EEC is connected to the torque motors for the control of the servo valves or to solenoids for the switching of hydraulic or air pressure. Each torque motor and solenoid has two electrical circuits, one for Channel A and one for Channel B. Individual harnesses are installed for each channel to connect these circuits to the EEC.

Both control software parts are operating at the same time to process the input data, but only one channel sends output signals for the control of actuators. This channel is called the channel in command. The other channel is the standby channel. This channel becomes the channel in command during the next engine start.

5.4.2.2 The Control Software

In both channels the same software is used. The software of a channel can be split into two basic parts according to their functions. These are:

- The control software
- The maintenance software

The basic structure of the EEC control software is similar to the system structure of the hydromechanical control unit. The main software sections are:

- Power management
- N1/EPR control
- Fuel metering valve control
- Subsystem control

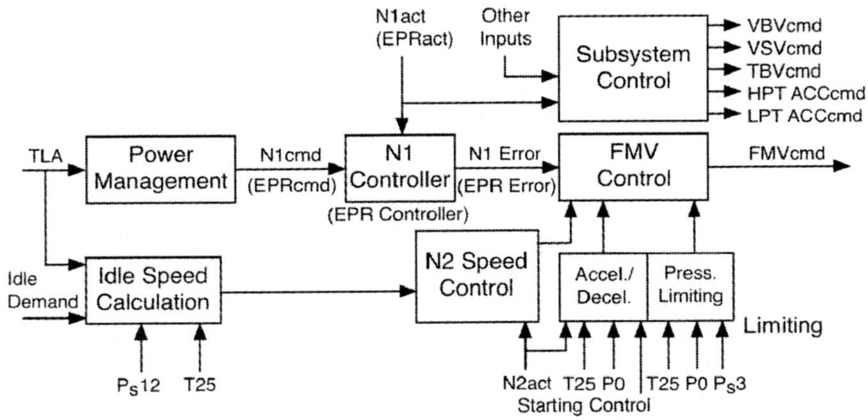

Fig. 5.21 Simplified structure of the control software based on the CFM56-5B software

Figure 5.21 shows a simplified block diagram of a control software. Dependent on the system design the power management section calculates the N1command or EPRcommand. Following the processing sequence for an N1 controlled engine, the next step is the N1 control. Here the actual N1 is compared to the N1command. If a difference is present, the N1 control section sends an acceleration or deceleration command to the fuel metering valve control section. This section calculates the FMV demand and sends it to the closed loop controller for the FMV. From the limiting section the FMV control section also receives inputs. The N2 speed control becomes active if the TLA is in the idle position and the current idle rating is based on N2 speed. The subsystem control section controls the actuators of the subsystems like VSV, VBV and ACC according to the changing operating parameters.

Most of the FADEC systems with the EPR as the thrust control parameter are able to operate in the N1 mode as an alternate mode. The EEC will switch to this mode if the sensing or calculation of the EPR is no longer possible. In the N1 mode the engine can be controlled manually similar to the normal EPR mode, but there may be some restrictions for autothrottle operation.

Through its outputs the EEC controls the actuators in a closed loop control circuit. For the necessary feedback information to the EEC a position sensor is installed in each actuator and connected to the EEC. Figure 5.22 shows this control principle with the related software sections within the EEC. The necessary position command comes from the control law of the respective function and is sent to the closed loop control section. Here the actual actuator position is compared to the commanded position. As long

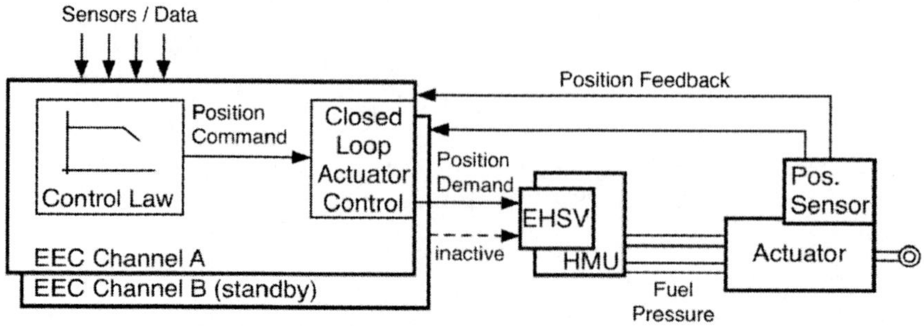

Fig. 5.22 The principle of closed loop actuator control with the electro hydraulic servo valve located in an HMU. The torque motor of the servo valve has one coil for each EEC channel

as an error exists an output current keeps the torque motor for the servo valve control out of the neutral position. Fuel pressure from the servo valve moves the actuator piston, which then again moves the position sensor. When the actuator reaches the commanded position, the closed loop control section stops the movement by setting the servo valve to the neutral position. For more details about servo valves see Appendix B.

The engines of an engine series are normally available with several thrust ratings. The control software is identical for all thrust ratings of an engine despite the information about the thrust rating of the individual engine. This information is stored in a plug, which is installed on the corresponding connector of the EEC. This plug is called rating plug or data entry plug. The thrust rating information is not the only information stored in this plug. All or some of the following can be stored additionally in the data entry plug:

- Engine serial number
- Fuel system configuration
- Combustor configuration
- Customized rating particularities
- Additional data for ECM
- N1/EPR modifier

This design allows an easy change of the thrust rating of an engine by changing the data entry plug. Through the data entry plug the EEC software is configured for the individual engine. Therefore the information stored in the plug belongs to the engine like the nameplate. To make this fact visible and to prevent a removal of the plug during EEC replacement, the plug is strapped to the engine case with a lanyard. The data entry plug data correspond to the data written on the engine nameplate. These data are the engine serial number, thrust rating and N1/EPR modifier.

Fig. 5.23 Data entry plug on the EEC of a CFM56-7B (© LTT)

The N1 or EPR modifier is used to correct production tolerances of the engine after the test cell run. To understand the principle of this correction for an N1 controlled engine, two brand new or freshly overhauled engines of the same type and thrust rating are compared; running on the same N1 under identical atmospheric conditions. The power management software will calculate the same N1command in both EECs. Due to production tolerances the two engines will generate different thrust forces that are normally higher than the nominal thrust. The purpose of the N1 modifier is to change the N1 command of the engines to a lower value, which leads to the same thrust on both engines. To accomplish this, an individual modifier level is stored in the data entry plug of each engine. With this modifier the EEC determines the N1 difference for the reduction of the calculated N1command. On the display for the pilots the unchanged N1command and actual N1 are shown. So the N1 indication of all engines are identical at the same thrust setting despite the different modifier levels.

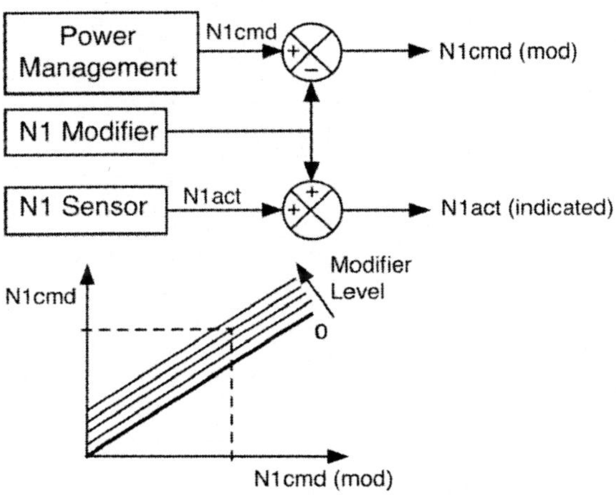

Fig. 5.24 Effect of the N1 modifier

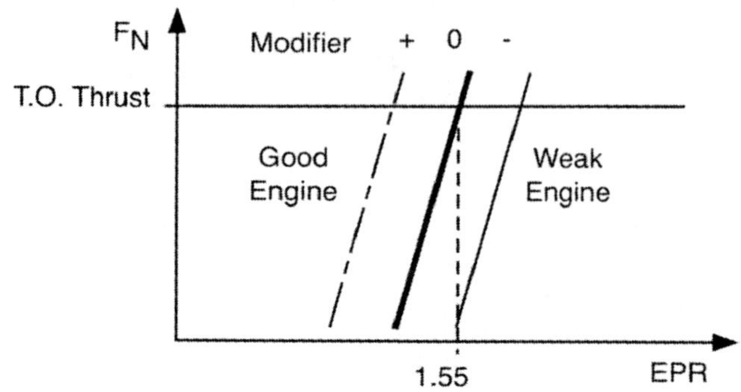

Fig. 5.25 Definition of EPR modifier. An engine that reaches the nominal thrust at an EPR of 1.55 would require an EPR modifier of 0

The function of an EPR modifier is slightly different. It assures that all engines produce their rated thrust at the same EPR. For this purpose the EPR modifier changes the sensed EPR to a lower or higher value. An engine that produces the nominal thrust at a higher than normal EPR for example, needs an EPR modifier that reduces the sensed EPR.

When an engine produces more thrust than the nominal thrust without a modifier, it will have a gain in EGT margin with the use of an N1 or EPR modifier. Both types of modifiers can only be specified after a test cell run because the necessary thrust measurement can be done in a test cell only.

5.4.2.3 The Maintenance Software

While the control software operates for engine control, the maintenance software operates parallel to it and monitors the whole system for failures. The detection of failures triggers fault messages to the central maintenance computer of the aircraft. At a higher fault class the fault messages also trigger messages in the ECAM or EICAS system. The maintenance software is the part of the software the mechanics communicate with via the central maintenance computer during troubleshooting. To make the trouble shooting as easy as possible, the maintenance software is able to isolate the most probable causes for a fault. With this information the isolation of the faulty component is facilitated. The fault state of the FADEC system is also used to calculate the dispatch level of the system before engine start. With this function the EEC can indicate if the system is unserviceable or the operating time until fault fixing is limited.

The maintenance software is also able to do system tests on ground with the engine not running. This makes a running engine for the tests unnecessary. The tests are used to verify a fault and for a system test after the replacement of a system component. The extent of the number of tests depends on the engine type. Common are mainly the electrical EEC test, the ignition test and the thrust reverser test.

The main task of the control software is to keep the engine running within its operating limit until the pilot shuts the engine down. This job has also to be done by the software if a fault or multiple faults occur. To make this possible the software is designed to be fault tolerant. This means it can operate the engine with lost input signals or lost functions. The latter occurs with multiple faults present. To always have the channel with the best health status as the channel in command, a managing software selects the channel in command depending on the health status. The health status is determined by the combination of present faults. The managing software changes the channel in command during the operation if the health status of the channel in command decreases below the health status of the standby channel.

The loss of an input signal does not lead directly to a change of the channel in command. If an input value from an aircraft system is lost, the EEC uses its own sensors to determine this value. If an input value from a sensor is lost, the EEC has several ways to react to keep the functions alive. It can use the input value from the other channel or it uses a synthesized value or it uses a default value. For important data all these possibilities are used in this order. For less important data only the use of a default value is possible. For the synthesizing of a lost input value the EEC calculates the value from several other sensed values on the engine. As an example the compressor inlet temperature can be calculated from shaft speed and compressor outlet temperature. This reaction to a sensor fault assures that the fault has no effect on the engine operation. The pilots see no change in engine behavior. Only in the maintenance report after the flight the sensor fault is visible.

Table 5.1 Fail-safe positions of some actuators on a CFM56-5B

COMPONENT	POSITION	IMPACT ON ENG. OPERATION
FMV	closed	0 kg/h
VSV	closed	engine operation impossible above idle
VBV	open	loss of max. T.O. power and/or high EGT
SAV	closed	no starter operation
HPTCC VALVE	closed 0%	9th stage air/clearances open
LPTCC VALVE	open 50%	clearances maintained
FUEL RETURN VALVE	closed	modulated idle (engine oil temperature controlled)

If a system function is lost due to an electric or hydraulic fault, the actuator for this function will move to its fail-safe position. In this state the affected system operates in a mode, which is uncritical for the engine, e.g. it can be kept within the operating limits. Depending on the lost function, such an operation with one or more systems in a fail-safe state can limit the engine operation. If the VBV system is in the fail-safe state for example, the VBVs will be open. The engine can't reach take-off thrust. Table 5.1 shows fail-safe positions of selected functions of a CFM56-5B.

5.4.3 The Sensors

As stated above, the sensors for the sensing of gas path values are located at the corresponding positions in the respective case. Additionally to the gas path values the EEC senses the shaft speeds and positions of several actuators and the thrust lever. The engine sensors are mainly designed as dual sensors. If an electrical sensor is not designed as an LRU for the replacement during line maintenance, a spare sensor is installed together with the main sensors for Channel A and Channel B.

Probes for total pressure or static pressure are installed to sense pressures. They are designed as single probes or are combined with a temperature sensor (P25 and T25 for example).

Temperatures are sensed with thermocouple type sensors or RTD type sensors (resistive thermal devices). Which type is used depends on the temperature range to be sensed. The thermocouple type sensors are mainly used at the HPC discharge and in the turbine area.

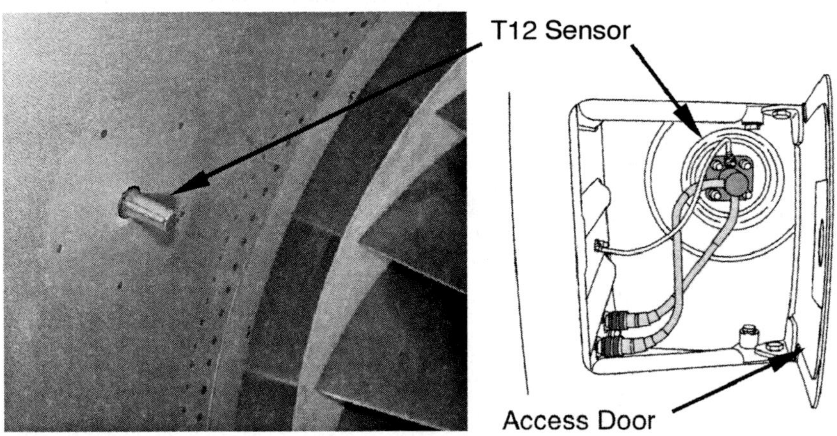

Fig. 5.26 The T12 sensor of the CFM56-7B installed in the air inlet in front of the fan rotor. The right picture shows the two electrical connections of this dual sensor below the access door in the outer skin of the air inlet (© CFM)

The shaft speed sensors are induction type sensors and operate without contact to the shaft. A toothed ring rotates near the sensing element and influences its magnetic field. For the sensing of the N1 speed the toothed ring is located on the N1 shaft within the front bearing compartment. The toothed ring or wheel for the N2 sensing (on twin-spool engines) is located in the accessory gearbox. To save the special N2 sensor, the alternate current of the FADEC alternator is used by some systems for the N2 sensing.

For the sensing of positions several types of sensors are used. The position sensor within a hydraulic actuator is an LVDT (linear variable differential transducer). This sensor senses any position the actuator piston can reach along the piston stroke. The positions of rotating valves are sensed with RVDTs (rotating variable differential transducer). On the thrust levers resolvers are used. For functions with only two states, switches are used to detect the actual position. These switches have dual contacts or two switches are installed to have separate circuits for the two channels.

5.4.4 Aircraft / Engine Interface

For the exchange of data between the EEC and the aircraft systems an interface between the engine and the aircraft systems is necessary. To minimize the number of wire strands used, most of the data are transmitted and received via data buses. The data buses from the EEC are routed to an

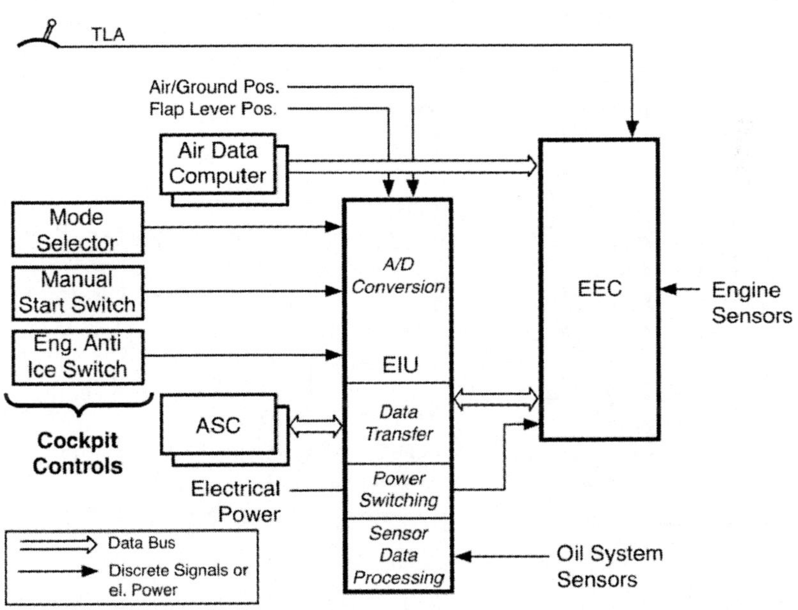

Fig. 5.27 Engine/aircraft interface of a CFM56-5B on an A320 (simplified)

interface computer within the airframe. In Airbus aircraft like A320 and A340 this computer is called engine interface unit (EIU) or engine interface and vibration monitoring unit (EIVMU, A340). One unit for each engine is installed. In a B737-800 the interface computers are the display electronic units (DEU). The interface functions may also be split and assigned to two different computers for each engine.

The number of data buses between the EEC and the interface computer is also limited to save weight, because the distance between the two computers is relatively long. The EEC in an A320 for example is connected via 5 data buses to the EIU. For a safe engine operation in the event of an EIU failure the data buses from the air data computers are direct connections between these computers and the EEC. The data the EEC receives from the air data computer are the Mach number, the total air pressure, the altitude and the total temperature. Other important aircraft data are the air/ground condition of the aircraft, flap lever position, bleed air demand and autothrust parameter for example. These data are transmitted by the respective aircraft systems computer (ASC) via the EIU.

The discrete inputs from the aircraft systems or engine controls are digitized by the interface computer. Usually these inputs are the switch positions of the engine control switches like mode selector, start switch, master switch and engine anti ice switch. The digital inputs from the various aircraft systems are transferred via the interface computer into the data buses to the EEC.

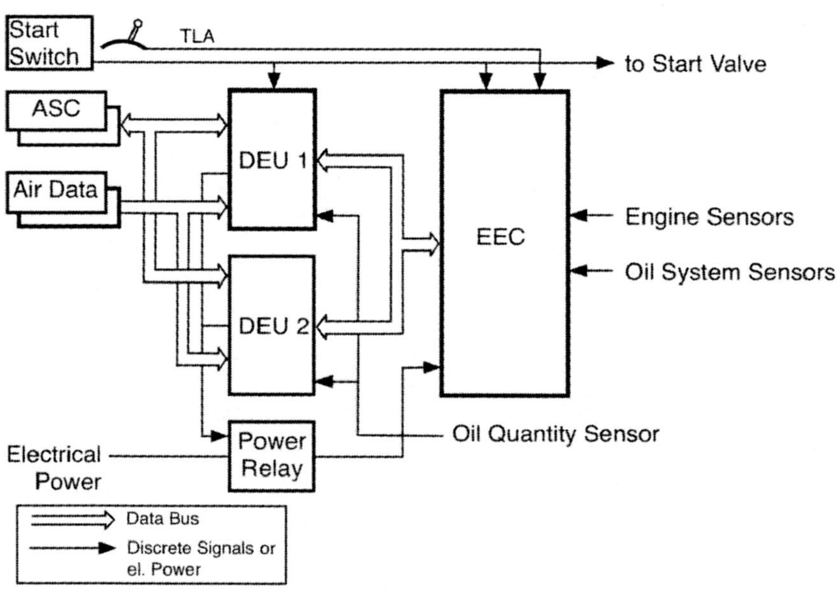

Fig. 5.28 Aircraft/engine interface of a CFM56-7B on a B737NG aircraft

In Airbus aircraft the power supply from the aircraft network for the EEC and the ignition system is also routed through the EIU. After engine shutdown on the ground the EIU disconnects the aircraft power from the EEC. In the B737-800 this disconnection is done by the DEU. This depowering assures a safe condition for the engines while the aircraft is parked with the electrical network powered.

Some functions of the engine oil system monitoring like oil quantity sensing and oil pressure sensing may be assigned to the interface computer. In these cases the interface computer is the data source for the indicating system. For all other engine data to be indicated in the cockpit, the EEC is the data source. In Airbus aircraft the data buses for the indication data are routed directly to the indication system computer. They are not routed via the EIUs to ensure the indication of engine data in the event of an EIU failure. In B737 aircraft all indication data from the EEC are sent to the DEUs, because the DEUs are part of the indicating system.

5.4.5 FADEC Systems of Selected Engines

In this section the FADEC systems of popular engines used in large numbers are presented to show the differences in the basic system designs. The different assignments of system functions to the system components can be recognized.

5.4.5.1 FADEC System of the CFM56-5B

The CFM56-5B is installed in aircraft of the A320 family. Its FADEC system has the basic design of GEAE FADEC systems. In these systems the servo valves are all located within the HMU. For an overview of this architecture see Fig. 5.29. The only electrical components in the actuators are the position sensors. The FADEC computer is called electronic control unit (ECU). In this system the same ECU is used as in the CF6-80C2 FADEC system with some improved details. It is from an earlier generation than the EEC of the CFM56-7B and it has no capability for the oil system indications. The oil system indication data are processed by the EIU. For the principle of the aircraft/engine interface see Fig. 5.27.

The main control parameter for thrust control in this system is the N1 speed. When the engine is controlled with the thrust lever, the ECU controls the thrust according to the TLA. The ECU senses the TLA with a sensor at the thrust lever. In the autothrust mode the ECU uses the N1cmd it receives from the auto flight system

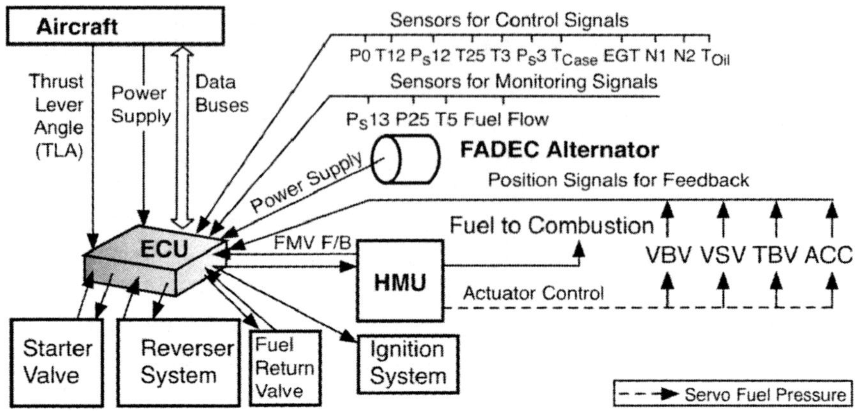

Fig. 5.29 Basic architecture of the FADEC system of a CFM56-5B

Additionally to the systems controlled for engine control the following systems are also controlled by the ECU:

- Ignition system
- Fuel return valve for heat management
- Starting system
- Reverser system

The engine sensors are arranged in two groups. One group contains the sensors for control signals, the other contains the sensors for the monitoring signals. Control signals are used for engine control purposes by the ECU and for engine condition monitoring (ECM). Monitoring signals are detected for ECM purposes only. The sensors for the monitoring signals are single sensors. All gas path sensors and the speed sensors are designed as LRUs.

Fig. 5.30 The ECU on the right side of the fan case of a CFM56-5B (© CFM)

Fig. 5.31 The HMU of the CFM56-5B (© CFM)

From the data entry plug the ECU receives the thrust rating, the actual engine configuration of the different possible configurations for a CFM56-5B and the N1 modifier.

The ability to control the ignition system and the starting system enables the FADEC system to execute automatic engine starts. During an automatic start the ECU monitors all relevant engine limits and prevents the exceedance of a limit by the interruption of the starting sequence. As an alternate procedure the manual engine start is also possible.

The ignition system is also activated by the ECU if a flameout is detected during engine operation. The necessary relays for the switching of the ignition power supply are installed within the ECU. The power supply for the ignition system comes from the aircraft network.

For the operation of the thrust reverser the ECU must detect the thrust lever angle in the reverse range and all the other limiting prerequisites (aircraft on ground, engine is running) must be fulfilled. During the reverser operation the ECU controls the engine operation in relation to the reverser operating condition. No other control components than the ECU and the thrust lever system are necessary for the operation of the reverser system.

If one or more faults are present in the FADEC system the fault messages released by the maintenance software are sent to the central maintenance computer in the aircraft. The maintenance software is also used for system tests. It can execute an electrical test of the ECU and an FADEC system test with engine motoring. During the latter test the system is supplied with fuel pressure from the fuel pump and the actuators can be moved during this test. This enables the testing of the feedback sensors.

To test the ignition system the ignition test is used. During the ignition test the ECU activates the ignition systems with the engine not running. Without the test function it would be necessary to deactivate other system components before the ignition system is activated manually for testing. The ECU monitors if electrical power is drawn by the igniter boxes. The operation of the igniter plugs must be checked audibly.

For the normal thrust reverser operation a running engine is necessary. The reverser test allows the operation of the reverser system with the engine not running (hydraulic pressure is provided by the electrical pumps). The movement of the reverser doors is controlled with the thrust levers.

5.4.5.2 FADEC System of the V2500-A5

The V2500-A5 is installed in the aircraft of the A320 family except the A318. The main functions and the main tasks of the system are the same as for the CFM56-5B. The aircraft/engine interface has the same design. The FADEC system of the V2500 is designed in the same architecture as the systems of Pratt & Whitney engines. In the hydromechanical part a fuel metering unit (FMU) is used. Thus the servo valves for the different actuators are located within the actuators. The supply with servo fuel for the actuators comes from the FMU inlet. Instead of a simple fuel return valve the more complex fuel diverter and return valve is used on the V2500 for the heat management of oil and fuel (see also Chapter 4.4). For this purpose an air-cooled oil cooler (ACOC) with a fuel pressure actuated air modulating valve is also installed. Fan air is used for the operation of this ACOC.

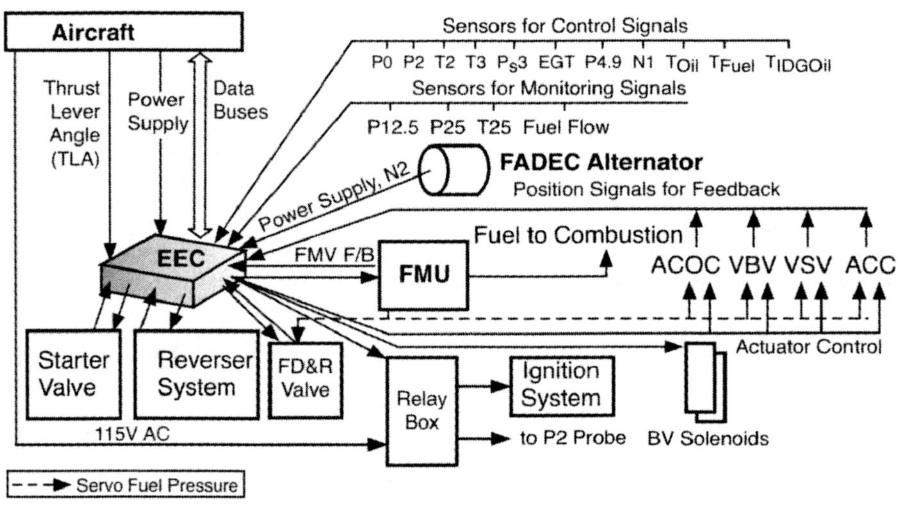

Fig. 5.32 Basic architecture of the FADEC system of a V2500-A5

Fig. 5.33 The EEC on a V2500-A5 (© LTT)

For the thrust control the EPR is used as the main control parameter. The actual EPR is calculated from P2 and P49. During autothrust operation the EEC receives the EPRcmd value from the auto flight system. If the EEC is not able to sense the pressures for the EPR calculation or the N1 mode switch is pressed by the pilot, the system operates in the alternate N1 mode. In this mode the N1 speed is the control parameter for thrust control and the engine can be operated over the full thrust range for all flight phases. The autothrust function cannot be used in this mode because the auto flight system sends EPRcmd values only and no N1cmd values.

The groups of engine sensors are shown in Fig. 5.32. To sense the N2 speed the EEC uses the frequency of the FADEC alternator. No special N2 sensor is installed. The sensors at the airflow station 25 and the N1 sensor are not designed as LRUs. They can only be replaced after the disassembly of the engine. The N1 sensor is installed in the front bearing compartment and has a spare probe, which can be connected to the EEC if one of the main probes fails. The connecting terminal of the probes is located at the rear face of the intermediate case.

The data entry plug from the EEC has stored the information about the thrust rating, the engine serial number and the EPR modifier.

A relay box installed in front of the EEC contains the relays for the switching of the ignition system. The P2 sensor installed in front of the fan is an electrically heated device. The heating is active at low air temperatures only. It is activated by the EEC. For this function the EEC uses a third relay in the relay box. The electrical power for the ignition system and for the probe heating is delivered from the aircraft.

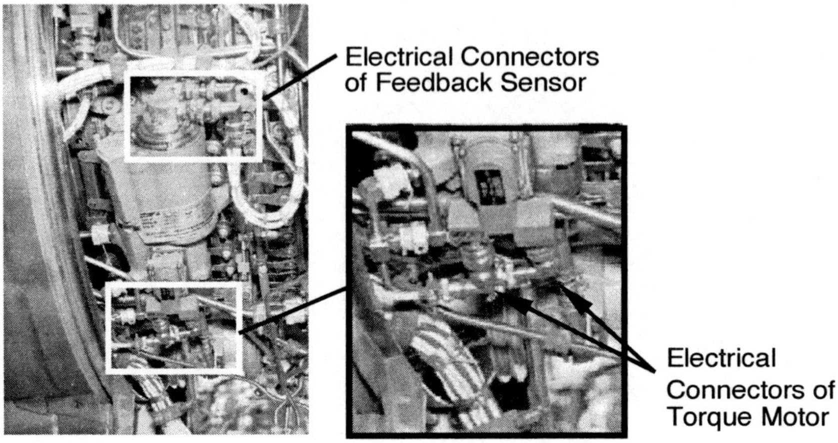

Fig. 5.34 VSV actuator of a V2500-A5 with its electrical connections (© LTT)

This engine has four pneumatically actuated bleed valves on the HPC. The actuating air pressure of these valves is switched on and off with the pneumatic solenoid valves. For each bleed valve one solenoid valve is installed and they are connected to the EEC.

The variety of system tests is very similar to that of the CFM56-5B but one essential test function is not available. This is the FADEC motoring test. While on the CFM56-5B the check of system components after its installation can be done with the FADEC motoring test, the same check on the V2500 can only be performed during an engine run-up. For the test of the P2 probe heating function a probe heater test is available.

5.4.5.3 FADEC System of the Trent 500

The Rolls-Royce Trent 500 is installed in the A340-500 and -600. The aircraft/engine interface is based on the design of the A320 but additional hardwired connections between the cockpit control switches and the EEC are installed. Due to this design the engine start functions can be used in the event of a failure of the interface computer. Each engine interface unit has an integrated vibration monitoring computer. This is the reason why these computers are called *engine interface and vibration monitoring unit* (EIVMU) in the A340. The sensors for the monitoring of the oil system are connected to the EEC.

In the FADEC system of the Trent 500 an HMU is used for the fuel metering functions and the VSV actuator control. The fuel system has no fuel return feature. For the cooling of the IDG oil an air cooled oil cooler is installed. For cooling control a 2-position air control valve is installed in the outlet duct of this cooler. This valve controls the flow of fan air through a

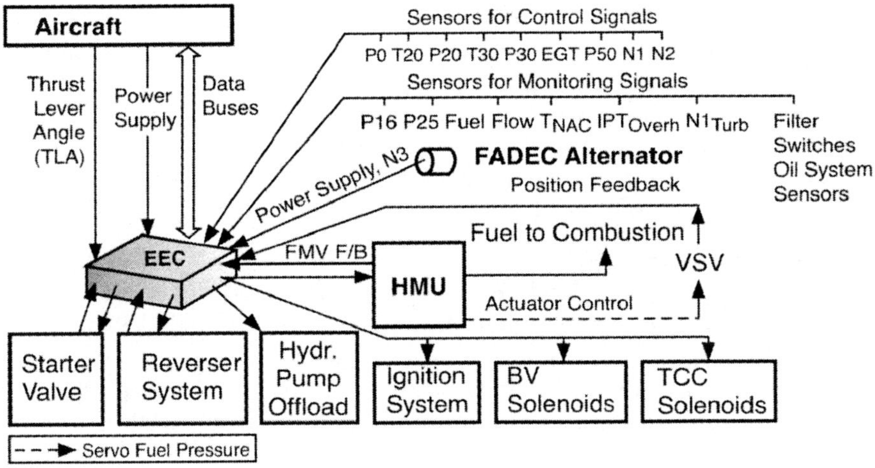

Fig. 5.35 Basic architecture of the FADEC system of a Trent 500

portion of the oil cooler. The airflow through the rest of the cooler is unrestricted. When the EEC commands the air control valve to open, the cooling effect increases.

All bleed valves and cooling valves of the engine air systems are operated pneumatically. For the control of these valves the EEC switches the respective solenoid valves electrically. Figure 5.35 shows the basic architecture of this FADEC system.

The EEC basically consists of the two channels A and B. The hardware of each channel can be divided into three groups, the power supply unit (power converter), the overspeed protection system and the control computer hardware. The power supply unit supplies the hardware of the overspeed protection system and of the control computer with electrical power. The overspeed protection system receives the inputs from the N1 and N2 speed probes of the compressor shafts and from the turbine N1 probe. This subsystem is independent of the operation of the control computer hardware. The additional N1 probe at the turbine is used for the detection of N1 shaft fractures. Such a fracture is detected by the comparison of the N1 values of the compressor shaft probes with the values of the turbine N1 probe. When the system detects an N1 or N2 overspeed condition or an N1 shaft fracture, it closes the overspeed valve inside the HMU. The shaft speed values are sent to the control computer to be processed and transmitted to the indication system. The N3 shaft speed is detected by using the alternate current of the FADEC alternator.

For thrust control this system also uses the EPR as the main control parameter. The EPR is calculated by the EEC on the basis of P20 and P50.

As the alternate operating mode the N1 mode is available. The EEC operates in this mode if the sensing of the pressures for the EPR determination is disturbed or the N1 mode switch is pressed by the pilot.

All the input data the EEC receives from the engine sensors are shown in Fig. 5.35. Very special sensors of the Trent engine are the IPT overheat sensors. These sensors sense the air temperatures in front of, and in the rear of, the intermediate pressure turbine (IPT) disk. The EEC sends a warning to the cockpit display if these temperatures exceed the limits. Figure 5.35 also shows the signal to the hydraulic pump offload solenoid. This function ensures that the hydraulic pump runs without load during windmill starts.

The data entry plug is installed on top of the EEC. From the data entry plug the EEC receives the following data:

- Engine serial number
- Thrust rating
- EPR trim
- EGT trim
- Engine configuration information

For system testing more test functions than have been described above are available. The ignition test and the P20/T20 probe heater test operate the same way as shown on the V2500.

During the VSV test the VSV actuators are moved while the engine is dry cranked by the starter. The VSV system is the only system with fuel pressure driven actuators and the EEC checks the actuator control components together with the position feedback sensors during this test.

The reverser test allows a check of the reverser operation as well as a test of the control valves in the hydraulic control unit of the system.

For the simulation of the engine running signal for the airframe systems the discrete output test is used. With this test it can be checked if the affected airframe systems are operating in the correct operating state after the start of the test.

During the hydraulic pump offload test the engine is dry cranked by the starter and the depressurization solenoid of the hydraulic pump is activated by the EEC. The operator must check the hydraulic pressure drop.

To test the electrically monitored chip detector a test is also available. It checks the operation of the chip detector and the state of chip accretion.

The pneumatically actuated handling bleed valves of the HPC can be checked with the bleed valve schedule test. This test is performed while the engine is running at idle speed on ground. The operator can open or close each handling bleed valve via the EEC menu during this test. This test is used to identify sticking valves and other faulty system components.

The next maintenance related item is the core washing procedure. With this function no test is started. It is used during the water wash of the engine and moves the VSV to the open position while the engine is cranked by the starter for washing.

5.4.6 Future Trends in FADEC System Designs

The basic design principles described above will also be used for the FADEC systems of the next generation engines. But these systems for engines under development today like the GEnX or comparable engines will have improved capabilities to make the engines more reliable and reduce the operating costs. GE Aviation sells these capabilities under the term "Intelligent Engine."

The fault isolation will be improved to get a complete on-board fault isolation by the system itself. The diagnostic capabilities will be improved for the gas path diagnostics as well as for the mechanical health diagnostics.

Due to the improved gas path diagnostics the systems will be able to determine the HPT deterioration and its effect on the EGT margin. This allows a real time adjustment of the active clearance control function to keep the efficiency of the HPT at its optimum. Another impact on the maintenance costs will have the capability to use the data of the real flight cycles for the life time calculation of the life limited parts (LLPs). This allows the optimum use of the LLPs because the effect of flight cycles with reduced temperatures can be taken into account for the life time calculations.

The oil systems will be monitored with a more sophisticated debris detection. The sensors are based on an inductive measurement technique which enables the system to detect, count and classify wear metal particles by size and type (ferromagnetic or non-ferromagnetic). This allows the system to determine the trend for the amount of particles in the oil.

The use of pressure sensors instead of pressure switches for oil filter monitoring makes a trend calculation for the filter contamination possible.

6 The Aircrew/Engine Interface

According to the law, the flight crew members are responsible for the proper operation of a flight and the safety of the airplane and its occupants. In order to fulfill these legal requirements the aircraft and the operating cockpit crew form a control loop in which the crew station provides the interface between the human controller and the aircraft. Concerning the engines this interface is accomplished with the engine indications and the engine controls. The following is a description of both system groups.

6.1 Engine Indications

An indicating system communicates data between the aircraft and the crew. The contents and format of the data must allow immediate interpretation of the operation of the aircraft and its systems.

If a system operation is automated, the crew's function becomes one of monitoring. But means of manual operation must always be provided and the consequences of a manual input immediately indicated.

The engines of an aircraft are undoubtedly critical for the operation and safety of a flight. Thus they play a dominant role in the indication and control system in the cockpit of an aircraft.

6.1.1 Indicated Engine Parameters

The pilots are responsible during aircraft operation that the engines and the engine systems are operated within their set limits. To monitor the engine with this objective and for the monitoring of the thrust output several engine parameters must be displayed to the pilots. These parameters can be grouped into three categories:

- Engine performance parameters
- Engine system parameters
- Mechanical engine condition parameters

The parameters to be indicated for the engine performance monitoring of a turbofan engine are:

- Engine Pressure Ratio (EPR)
- Rotor Speeds N1, N2, N3
- Exhaust Gas Temperature (EGT)
- Fuel Flow (FF)

The indications of these parameters are known as the main engine indications. As stated before these parameters are indicated for the monitoring of the engine thrust setting and also for the monitoring of the engine operating limits set for rotor speeds, gas temperature and thrust output.

6.1.1.1 Thrust Indicating Parameters

According to the certification standard EASA CS 25.1305 (and FAR 25.1305) "Power Plant Instruments" nearly all of these indications are mandatory except the EPR indication. This parameter is used for the control of the thrust output only. In conformance with CS 25.1305 a thrust indication or the indication of a parameter equivalent to the thrust is mandatory. The respective sub-paragraph of CS 25.1305 reads as follows:

An indicator to indicate thrust, or a parameter that is directly related to thrust, to the pilot. The indication must be based on the direct measurement of thrust or of the parameters that are directly related to thrust. The indicator must indicate a change in thrust resulting from any engine malfunction, damage or deterioration.

Thus the N1 indication also fulfills this requirement because the fan moves the total airflow of the engine. Each engine type has an individual relationship between the fan speed, the total airflow and the thrust. Therefore the N1 speed is directly related to the generated thrust.

If the engine manufacturer intends the use of the EPR for an engine design, its indication is provided in the aircraft. In this case the EPR is the parameter used to express the engine thrust setting. Because the EPR is a ratio of two pressures, no unit is assigned to it.

6.1.1.2 Shaft Speeds

The shaft speeds N1, N2 and, if applicable, N3 are indicated in percent of the 100% shaft speed of the respective rotor system. The 100% shaft speeds are determined by the designers during the engine design phase. They need not match with the maximum permissible shaft speeds (redline speeds) of the real engine. Some engines have redline speeds of more than

100%. During the design phase of these engines the physical shaft speed range was extended to higher values and the earlier determined 100% shaft speeds were kept on the same physical speeds.

Additionally to the monitoring of the shaft speed limits the N1 speed is often used as the equivalent thrust indication parameter. In this case the N1 indication additionally provides information of the necessary N1 for a required thrust setting and a thrust lever position indication to facilitate the manual setting of thrust. The speed of the HP rotor (N2 or N3) is used for the monitoring of the idle speeds because the idle ratings are mainly based on HP rotor speeds. The HP rotor speed is also important for the monitoring of the engine start sequence because the steps of the sequence refer to N2 or N3 values.

6.1.1.3 Exhaust Gas Temperature

A very important limit for the engine is the gas temperature limit because an exceedance of this limit can lead to immediate damages of engine parts or a life reduction of engine parts. If a temperature limit exceedance were undetected it could lead to later part failure due to reduced part life. The calculated life times of the life limited parts (LLPs) are based on an engine operation within the temperature limits.

For the monitoring of the engine gas temperature it is sensed on a position in the gas path where it is technically possible due to the local temperature. This position is usually located behind the HPT or behind the LPT. The temperature sensed at one of these locations is defined as the engine's *Exhaust Gas Temperature (EGT)* by the designer. Which position the designer selects for the installation of the EGT sensors depends mainly on the engine manufacturer's philosophy. The most important temperature to be limited in the hot section of the gas path is *the Turbine Inlet Temperature (TIT)*. But due to the high temperature level at the turbine inlet this temperature cannot be sensed directly. The temperature sensors do not have the capability of sensing such a high temperature.

The gas temperature limits for an engine are set for take-off thrust setting and for maximum continuous thrust setting. They are expressed in EGT values. If the EGT is kept below these limits the TIT is also kept below its limits. The EGT is indicated in °C or °F.

6.1.1.4 Fuel Flow

The *fuel flow* is used to monitor the actual fuel consumption of the engine. Together with the other main parameters the fuel flow can be used to analyze the engine's efficiency. The total fuel consumed by an engine is indicated as "fuel used". This parameter is calculated from the fuel flow of the respective engine since engine start. During flight the total fuel consumption of all engines is used to calculate the present fuel on board (FOB). The fuel flow is indicated in kg/h or lb/h.

6.1.1.5 Engine Systems Indications

The indicated engine systems parameters are used for the monitoring of the systems conditions and for the monitoring of the systems limits. The only indication of the engine fuel system is often the fuel filter clogging indication. The fuel flow indication is also an indication of the engine fuel system but it is assigned to the main engine indications. The other engine system indications are related to the lubrication (or oil) system and the reverser system.

The indications mandatory for oil system parameters with the mostly used units are:

- Oil quantity (in quarts or percent)
- Oil pressure (in psi)
- Oil temperature (in °C or °F)

Warnings and cautions must be provided for:

- Low oil pressure (warning)
- Oil filter clogging (caution)

By interpreting each oil system indication in relation to each other, abnormal oil system conditions can be detected.

The reverser system indication consists of a reverser position indication (reverser unlocked light) on the center instrument panel and a warning indication for this system (reverser warning light).

For the monitoring of the mechanical condition of the rotor systems a vibration indication is provided. The vibration indication shows the vibration induced by imbalance. When compressor blades are damaged by foreign objects (FOD), contaminated with accumulated dirt or the lubrication of the fan blade roots has lost its effect, the vibration of the respective rotor system increases. Ice accretion on the compressor blades also leads to a higher vibration level.

The vibration is indicated for each rotor system of an engine. The vibration monitoring computer receives the signals from the vibration sensors and the rotor speeds as the inputs for the processing. It filters the sensor signals having the shaft frequencies out of the broadband sensor signals. Thus it uses one sensor value of each rotor for the calculation of the indicating data of each rotor. For this indication no unit is used. The indicated values are related to the load the vibration causes to the engine parts.

Other indications required by CS 25.1305 are related to fire warning, engine anti ice and the aircraft fuel system. These systems are not part of the engine systems and not discussed in this chapter.

6.1.2 Engine Indications on an Electronic Instrument System

6.1.2.1 The Airbus ECAM System

The indication of engine parameters by an Electronic Instrument System (EIS) will be explained by using the Airbus philosophy. The Airbus aircraft communicates to the crew through its EIS, which is divided into 2 subsystems:

- Electronic Flight Instrument System (EFIS)
- Electronic Centralized Aircraft Monitor (ECAM)

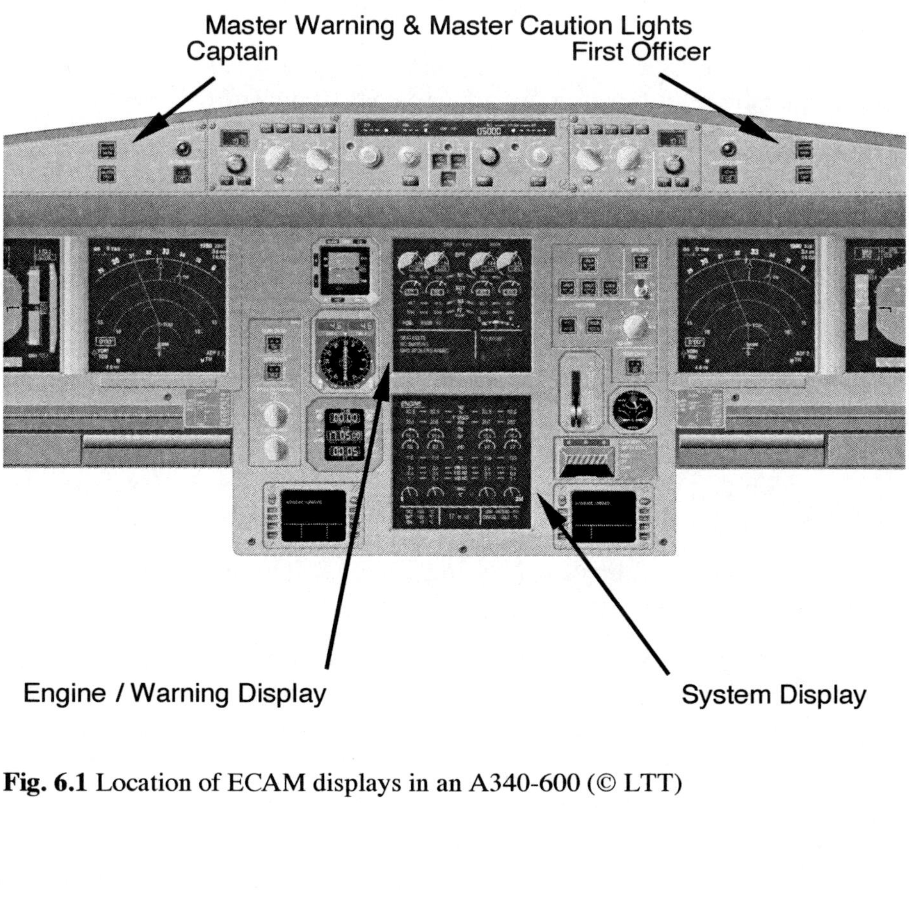

Fig. 6.1 Location of ECAM displays in an A340-600 (© LTT)

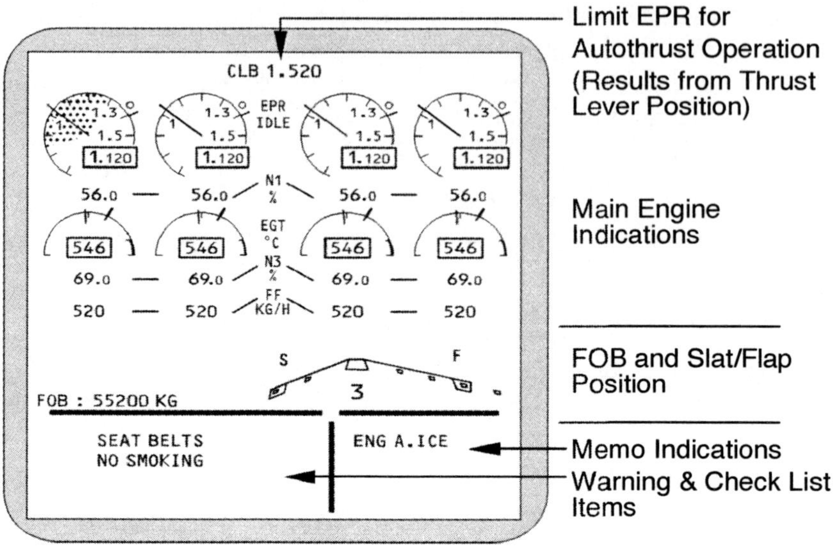

Fig. 6.2 Engine/Warning Display of an A340-600

In addition there are the following means of turning the crew's attention to the indication if a function diverts or data drift away from normal:

- Aural warnings
- Master warning lights
- Master caution lights
- Fault lights

The aircraft system indications, including the engine indications, are displayed by the ECAM system. The ECAM system provides information via two displays. They are located in the center of the forward instrument panel as shown in Fig.6.1. The two displays are:

- The Engine/Warning Display (E/WD), normally on the upper ECAM display unit
- The System Display (SD), normally on the lower display unit

Figure 6.2 shows the Engine/Warning Display. It provides the following information continuously:

- Basic engine parameters
  - EPR (if applicable)
  - N1
  - EGT

  - N2, (N3)
  - Fuel Flow
- Slat and flap configuration
- Messages and interactive checklists for normal and abnormal operation
- Total fuel on board

The indication of the main parameter used for thrust setting (power management parameter) provides some additional features to facilitate the manual setting of the thrust and the monitoring of the thrust changes performed by the flight management system. The EPR indication in Fig. 6.3 shows these features. If the N1 speed is used for thrust setting, the same features are displayed on the N1 indication. In case of an EPR indication these are:

- EPR Command
  - the green needle corresponds to EPR demanded by FADEC
  - only displayed when Autothrust is active
- EPR Transient
  - symbolizes the difference between EPR demanded by FADEC and actual EPR
  - only displayed when Autothrust is active
- EPR Trend
  - is displayed as a green triangle next to the EPR COMMAND needle for indication of the EPR tendency
  - only displayed when Autothrust is active

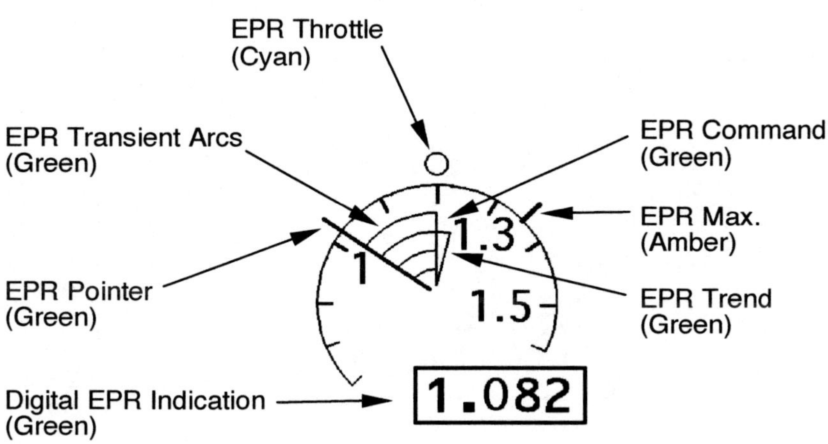

Fig. 6.3 EPR Indication with thrust setting facilities

- EPR Throttle
  - shows the EPR value the engine reaches with the current thrust lever position
- EPR Maximum
  - the EPR max is displayed by means of a thick amber mark across the EPR scale
  - corresponds to the EPR limit of the full forward position of the thrust levers(TOGA mode)
  - not displayed in reverse mode or when the engine is off

The position indication of the thrust reverser is shown in the EPR scale when the reverser is deployed or when it is accidentally unlocked in flight. Figure 6.4 shows the EPR scale with the reverser position indication. This symbol can indicate three system states of the reverser:

- REV symbol amber
  - Reverser is selected and moving
- REV symbol green
  - Reverser is selected and deployed
- REV symbol amber flashing
  - Reverser is unlocked while forward thrust is selected

Thus the upper ECAM display allows the pilots to continuously monitor basic parameters of the engines. The lower ECAM display provides simplified diagrams of the major aircraft systems displayed as so-called system pages. These system pages are displayed automatically in a flight phase controlled sequence. This automatic sequence can be overridden by a manual selection of a page on the ECAM control panel.

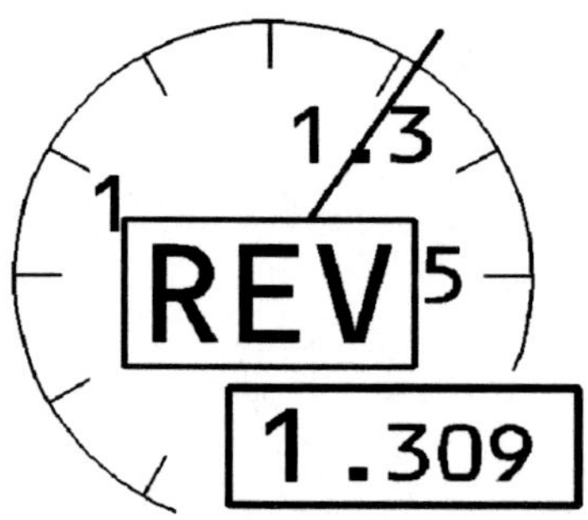

Fig. 6.4 Reverser position indication within the EPR scale

The indications of the engine systems are shown on the engine system page (see Fig. 6.5). The parameters shown on this page are:

- Fuel used
- Oil quantity
- Oil pressure
- Oil temperature
- Nacelle temperature (exchanged with start valve position, air pressure and ignition indication during engine start)
- Vibration level
- Oil filter and fuel filter clog symbols

By selecting the appropriate system page the pilot will find information about those systems that are powered by the engine, or that supply the engine. These are:

- Pneumatic system
- Electric system
- Hydraulic system
- Fuel system

The electronic indicating system is much more than a simple indicating system: The ECAM also monitors the aircraft systems and provides instructions for actions to be taken by the crew during normal and abnormal operations.

To ensure this a flight warning computer (FWC) and a system data acquisition computer (SDAC) generate alert messages and memo information for display on the upper ECAM display. Furthermore they trigger:

- Aural alerts
- Synthetic voice messages
- Warning and caution lights

When a failure is detected, the E/WD displays the failure message, instructs which actions have to be taken and indicates which systems are affected by this failure. The system display shows the appropriate system page related to this fault automatically.

Thus the display of the engine parameters is integrated into a centralized indicating and monitoring system. The engines play a dominant role in the display system with primary parameters continuously available on the upper display and additional parameters selectable on the lower display.

The system furthermore continuously monitors the engine operating parameters. The crew is advised when a parameter shift has occurred. In case of an engine failure the crew is alerted, provided with instructions what actions have to be taken and also informed which systems are affected.

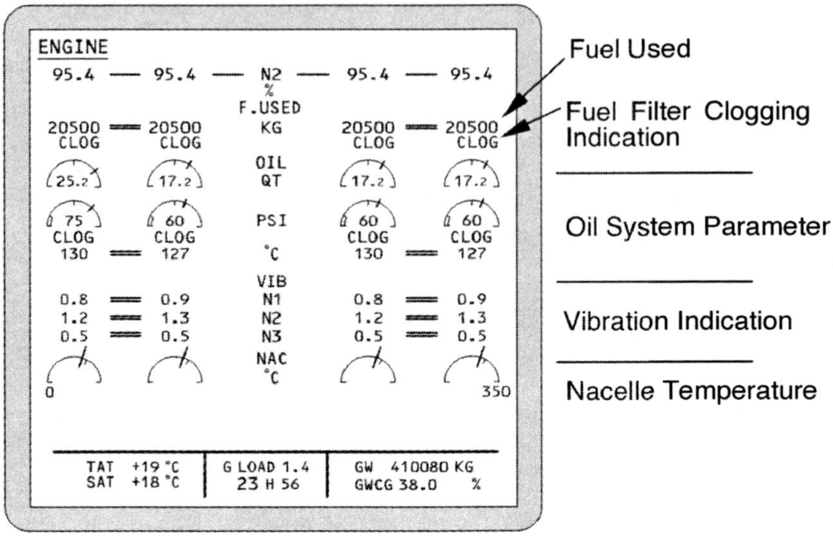

Fig. 6.5 System Display of an A430-600

A monitoring system like the ECAM system reduces the workload of the flight crew for normal system monitoring and much more for the management of a system fault. This is achieved not only by alerting the pilots in case of a failure but especially by telling them what to do to achieve the safest configuration of the remaining normal operating systems.

6.1.2.2 Engine Data for the ECAM System

For the description of the data transmissions for the engine indications on the ECAM system the system of the A340-600 is used as an example. A simplified system schematic of this system is shown in Fig. 6.6.

For the operation of the ECAM system the following computers with different functions are installed:

- Display Management Computers (DMC) for the generation of the graphics on the displays
- Flight Warning Computers (FWC) for parameter monitoring and generation of warnings
- System Data Acquisition Computers (SDAC) for aircraft systems parameter monitoring and generation of cautions (the SDACs receive no inputs from the engine systems)

For the indication of engine data and the generation of warnings and cautions the FWCs and the DMCs receive the necessary data from the computers connected to the engine sensors. These are mainly the electronic engine control (EEC) and the engine interface and vibration monitoring unit (EIVMU). The only sensor connected to another computer is the low oil pressure switch. It is connected to the FWC and its purpose is to trigger the low oil pressure warning. This warning is generated by the FWC and brought to display by the DMCs.

The parameters sensed by the EEC are sent via the data buses to the DMCs and to the FWCs. The data of the main engine parameters EPR, N1, N2, N3, EGT and fuel flow are sent directly to the DMCs and FWCs. The other parameters are sent by the EEC via the data bus connections of the EIVMU.

The vibration sensors of the engine are connected directly to the EIVMU because this computer does the data processing necessary for the vibration indication. For this data processing the shaft speeds are also input signals. The shaft speed data are transmitted by the EEC to the EIVMU.

The vibration monitoring computer integrated into the EIVMU has more functions than processing indicating data only. This computer is able to

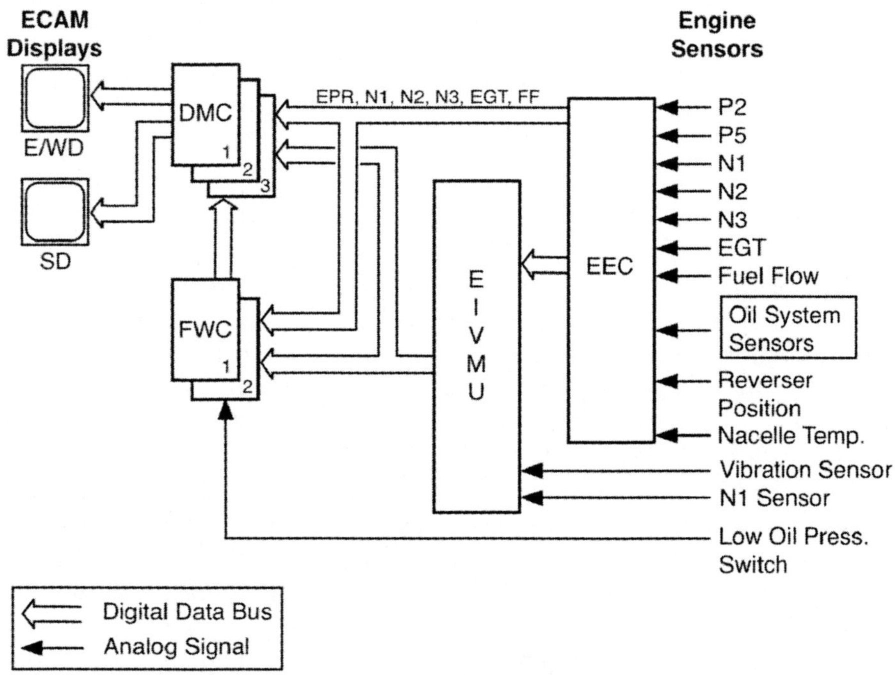

Fig. 6.6 ECAM system schematic for the engine indications of an A340-600 (simplified)

calculate additionally to the vibration value the position of the imbalance on the fan rotor. This is the input for the ground-based calculations to determine balance weight size and position for trim balancing if the vibration level exceeds a limit set by the maintenance.

A special N1 sensor is installed for this function in the engine and connected directly to the EIVMU. It is also shown in the ECAM schematic. Its phase signal has a defined position on the shaft relative to the fan blade No.1, which is the reference fan blade for the phase angle of an unbalance.

6.1.3 Engine Indications on a Classical Instrument System

6.1.3.1 Cockpit Instruments

For the description of a classical engine indication system the indication system of the B737-300 is used as an example. For this aircraft the engine and hydraulic system indications on the center instrument panel were available with individual instruments or with two large LCD instrument modules. In the following the variant with individual instruments is shown.

For the indication of the main engine parameters of the CFM56-3 this system consists of one instrument for:

- N1
- EGT
- N2
- Fuel Flow

The instruments for these parameters are installed in two columns, one for each engine. To the right of these columns the instruments for the oil system indications (pressure, temperature and quantity) and vibration indications of the two engines are located. Figure 6.7 shows this arrangement.

The main parameter used for thrust setting is the N1 speed. Thus in the N1 indicator a window for the digital indication of the N1command is provided. The N1 value indicated in this window is calculated by the flight management computer. A bug positioned on the analog scale indicates this value in analog form. For the manual setting of the N1command a setting knob on the front face of the indicator is provided. Figure 6.8 shows these features in detail.

The fuel used value is displayed digitally in the scale window within the fuel flow indicator scale. This value is reset to zero manually with a pushbutton on the instrument panel before the engines are started. The indicator itself calculates the fuel used value.

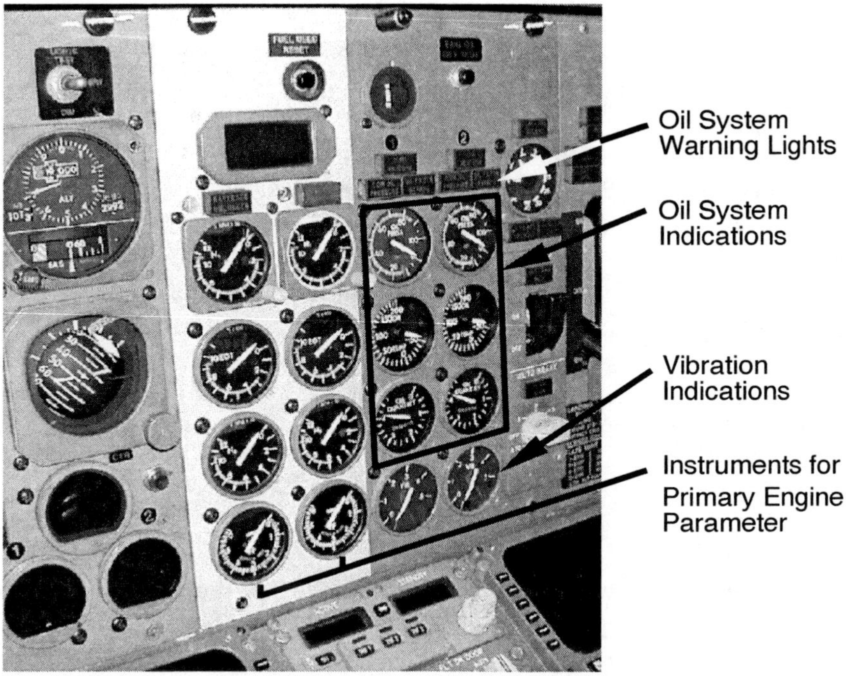

Fig. 6.7 Engine indications on the center instrument panel of a B737-300 (© LTT)

Above the instrument columns of the main parameters the reverser-unlocked lights are installed for the reverser position indication and above the oil system instruments the warning lights for low oil pressure and oil filter clogging are installed. Above the latter two lights the respective start-valve-open-light for an engine is installed. The fuel filter clogging light is installed on the fuel system section of the overhead panel.

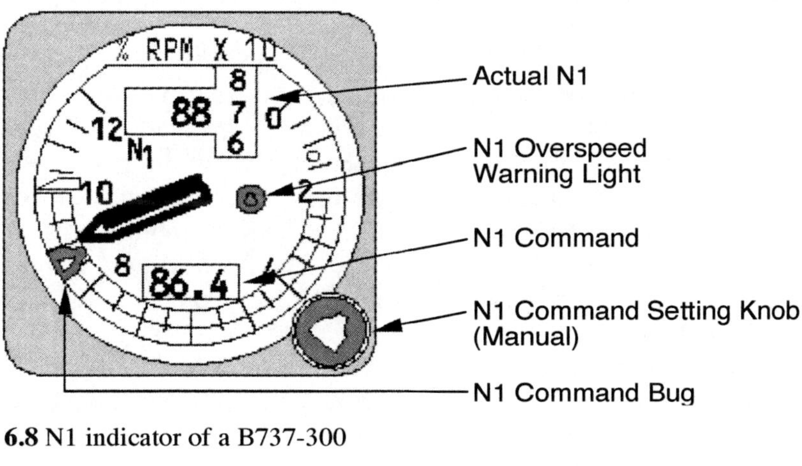

Fig. 6.8 N1 indicator of a B737-300

When a warning light is triggered, the master caution lights on the glare shield in front of the pilots also light up to turn the pilot's attention to the activated warning. If a limit exceedance of N1, EGT or N2 occurs, a red light in the respective instrument scale lights up.

In such a system one warning light for each warning is necessary. The illuminated warning light alerts the pilots but there is no more information by the system what to do to ensure a safe operation. This information is provided by checklists that have to be read by the pilots before any change in the system operating configurations is made. Compared to the electronic instrument system the pilots need more time for a correct reaction to a warning. Due to the more complex systems in modern aircraft and engines the intention of the designers is to reduce the workload of the pilots when a system malfunction occurs. Thus the electronic instrument system is used in all modern transport aircraft in production today.

6.1.2.3 Indicating System

The architecture of the indicating system for the engine parameters is very simple. The indicators installed in the instrument panel contain a tiny computer for the determination of the value to be sensed. This computer drives the instrument's pointer and the digital display. The power supply for the indicator comes from the aircraft network. If a sensor requires power supply (like the oil pressure sensor), it is supplied together with the indicator from the aircraft. Figure 6.9 shows the schematic of this system without the power supply for simplification.

To make the measurement possible, each indicator is directly connected to its assigned sensor on the engine. The warning lights for low oil pressure, fuel filter clogging and oil filter clogging are activated by the assigned switches on the engine. Each switch connects its warning light to ground when it closes at its switching point.

For the vibration indication the design is slightly different. The vibration sensors of the engine are connected to the Airborne Vibration Monitoring Signal Conditioner (AVM Signal Conditioner). This computer is also connected to the engine speed sensors for the sensing of the rotor speeds. It processes the sensed data and calculates the indication data. It drives the vibration indicator for the indication of the calculated vibration value. The vibration indicators have only one pointer. The AVM Signal Conditioner calculates four vibration values (one value for each shaft from the inputs of the two vibration sensors). The highest of these values is indicated by the indicator. For the flight crew it is not visible which rotor (N1 or N2) the indicated value belongs to. This can be revealed after landing by a data readout of the AVM Signal Conditioner done by the maintenance crew.

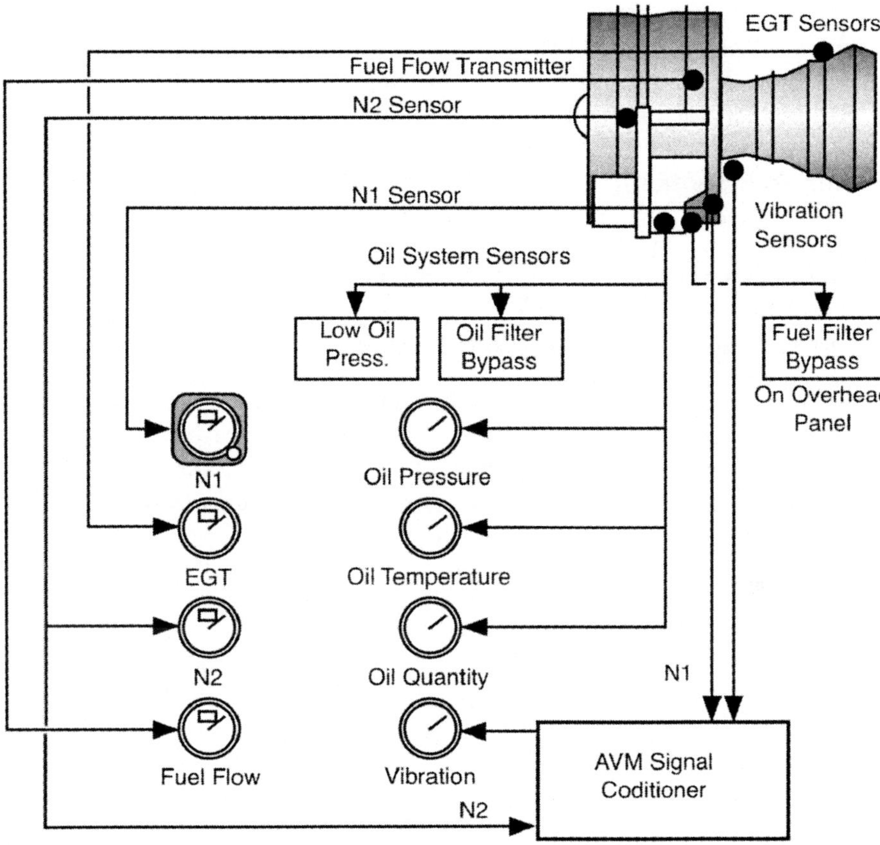

Fig. 6.9 Engine indicating system for one engine of a B737-300 (simplified)

The AVM Signal Conditioner stores all four highest vibration values of a flight in its memory. It can store the vibration data and fault data for 32 flights. One AVM Signal Conditioner is installed in the aircraft for both engines.

6.2 Engine Controls

This subchapter contains the description of the systems for the manual and automatic control of the engine. For the manual thrust control the airworthiness standards require one thrust lever for each engine. The thrust levers must be separately movable and all together with one hand.

Forward thrust control and reverse thrust control must be distinguished from each other. The reverse thrust controls are also used for the control of the thrust reverser system. The transmission for the thrust lever angle between the thrust lever and the fuel control system also belongs to the thrust lever system. This transmission is either mechanical or electrical.

For the control of the HP fuel shut-off valve within the MEC or the HMU a start lever or a master switch is installed in the cockpit. For this system the signal transmission can also be either mechanical or electrical.

6.2.1 HP Fuel Shut-Off Control System

6.2.1.1 Mechanical HP Shut-off Valve Control

If a hydromechanical fuel control system is installed on the engine, one used design variant is the mechanical start lever system. In this system the start lever, installed behind the thrust lever, has a mechanical connection to the shut-off lever of the FCU. This mechanical connection can be a typical cable system or a push-pull cable over the whole length.

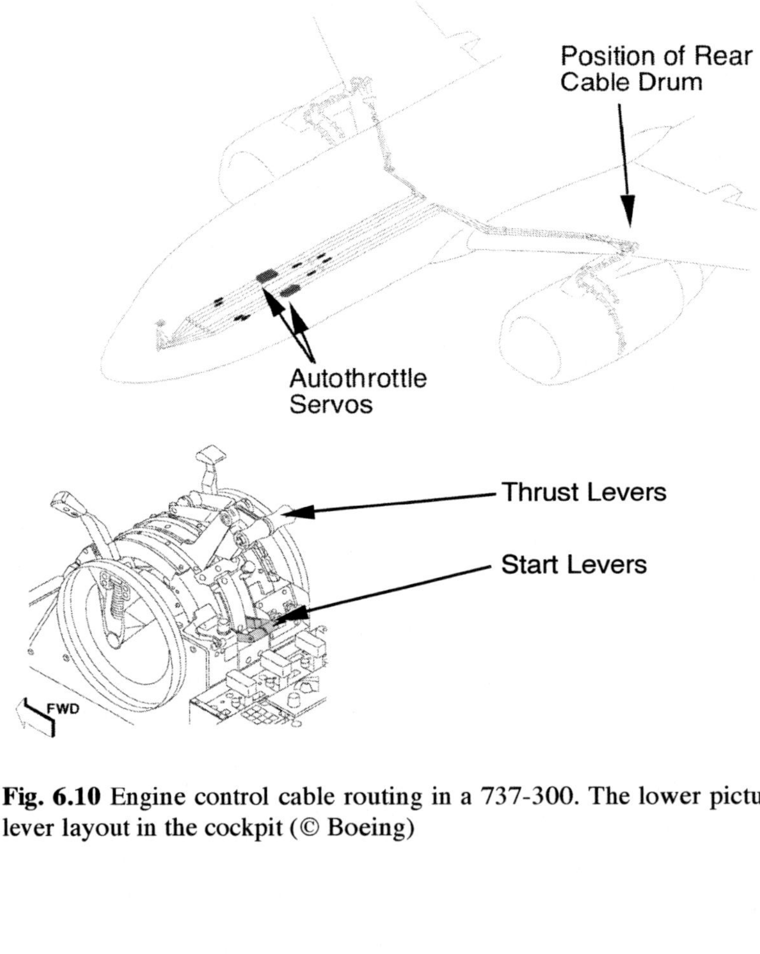

Fig. 6.10 Engine control cable routing in a 737-300. The lower picture shows the lever layout in the cockpit (© Boeing)

If a cable system is used, the start lever is connected to a cable drum installed below the start lever. From this drum the two cables are running down the fuselage to the engine pylons. Here they end at a second cable drum. From this drum a short push-pull cable transmits the movement to the shut-off lever of the FCU. See Fig. 6.10.

The start lever also operates the switches for the ignition systems of the engine in parallel. The switches are installed beside the forward cable drum and are closed by cams on the drum when the start lever is moved to the idle (or on) position. Additionally to the ignition switches start lever actuated system switches are installed. They activate airframe systems functions depending on the running state of the engine and they allow the opening of the low pressure fuel shut-off valve. This valve is installed between the aircraft fuel tank and the engine. Figure 6.11 shows the principle of the start lever system of a B737-300.

6.2.1.2 Electrical HP Shut-off Valve Control

The more sophisticated design for a HP shut-off valve control system is the electrically controlled system. In this system the operation of the HP shut-off valve is controlled with a solenoid valve installed on the FCU, or the HMU (FMU) if a FADEC system is used. The power for the solenoid valve is controlled in the cockpit by the use of a switch instead of the start lever. This switch is called the engine master switch (Airbus) or the fuel control switch (Boeing). An exception is the design in the B737NG. Here

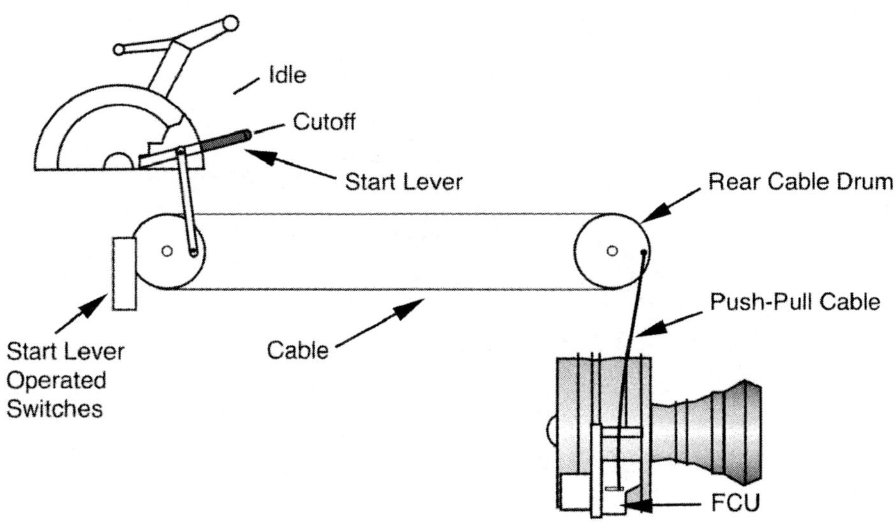

Fig. 6.11 Mechanical start lever system (simplified)

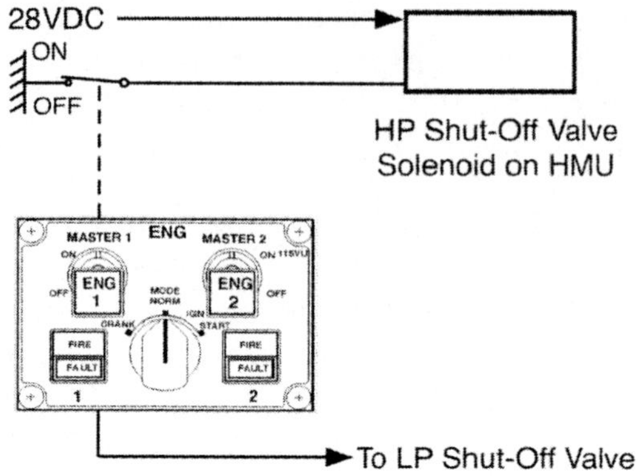

Fig. 6.12 HP shut-off valve control system of an A320 (simplified)

the start lever design from its non-FADEC predecessor is used. The start levers actuate the electrical switches. The solenoid valve in all designs is energized for the closing of the HP shut-off valve and deenergized during engine operation. In Fig. 6.12 the HP shut-off valve control system of the A320 shows this detail.

To prevent the operation of a wrong master switch during an engine fire alarm, an additional fire warning light is installed in each master switch or in the panel behind the switch. This fire warning light comes on together with the fire warning of the corresponding engine.

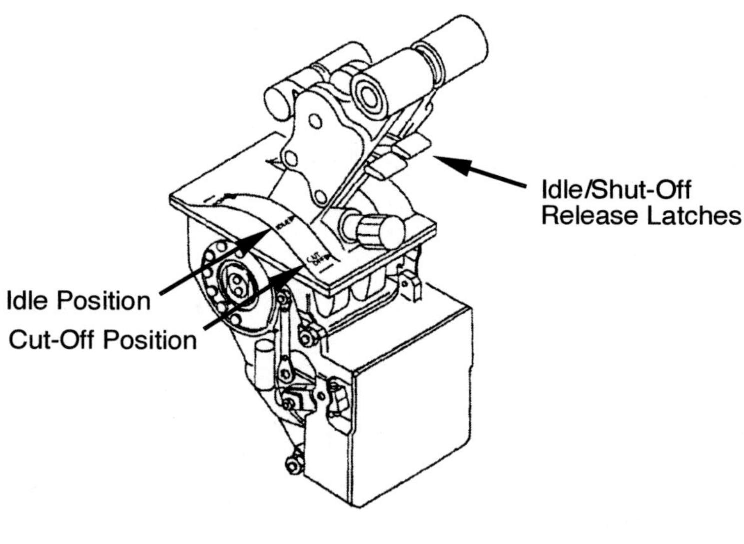

Fig. 6.13 The single lever control system of a CRJ200 (© LTT)

6.2.1.3 HP Shut-off Valve Control with Single Lever Control Systems

In business aircraft from the small jet up to the Bombardier regional jets the function of the start lever is often incorporated into the thrust lever. Such a control system is called a single lever control system. In this system design the thrust lever has a second rear stop behind the idle stop, the cut-off position. During the engine start sequence the thrust lever is moved from the cut-off position to the idle position to release the fuel flow. To move the thrust lever to the shut-off position during engine shutdown, a latch installed in the thrust lever must be released. Figure 6.13 shows the thrust lever layout of a CRJ200.

6.2.2 Mechanical Thrust Lever Systems

6.2.2.1 Forward Thrust Control

In mechanical thrust lever systems the change of the thrust lever angle during the movement of the lever is transmitted to the FCU via a cable system or with a single push-pull cable. In aircraft with wing mounted engines the cable system usually ends in the wing leading edge above the pylon. From there a push-pull cable transmits the lever movement to the FCU power lever on the engine.

The thrust levers themselves are installed in the center console of the cockpit. An adjustable friction brake located on the axle of the levers causes sufficient friction to keep the levers in the selected position. The friction prevents unintentional and rapid lever movements caused by the

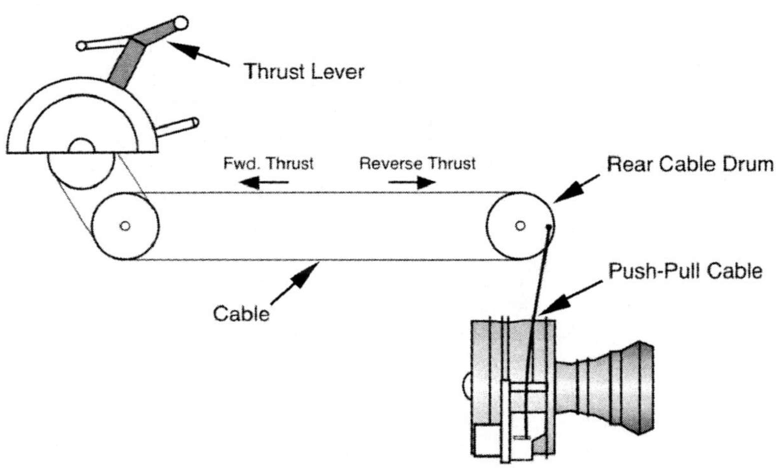

Fig. 6.14 Mechanical thrust lever system (simplified)

pilot because it creates an artificial feel. It also prevents the autonomous lever movement caused by vibration. The friction brake can also be located in the autothrottle servos.

In a two-lever system of transport aircraft the rear stop of the thrust lever is the idle position. The forward stop is designed at the maximum required thrust lever angle for take-off thrust. If the engine operates with a supervisory control unit it is usually not necessary to move the thrust lever to the forward stop for the selection of take-off thrust. This may only be necessary in the event of a failure of the supervisory control unit.

In aircraft equipped with thrust reversers the reverse thrust levers are installed on the thrust levers. A mechanical latch system allows the movement of the reverse levers only in the idle position of the forward thrust levers. If a reverse lever is moved away from the stow position, the movement of the forward thrust lever is prevented by a built-in mechanical lock.

6.2.2.2 Reverse Thrust Control

For the use of reverse thrust the thrust reverser must be deployed into the reverse position and the engine power for the required thrust must be selected. The movement of the thrust reverser from the stow position into the reverse position is triggered by control switches in the thrust lever assembly or beside the forward cable drum when the reverse lever is pulled away

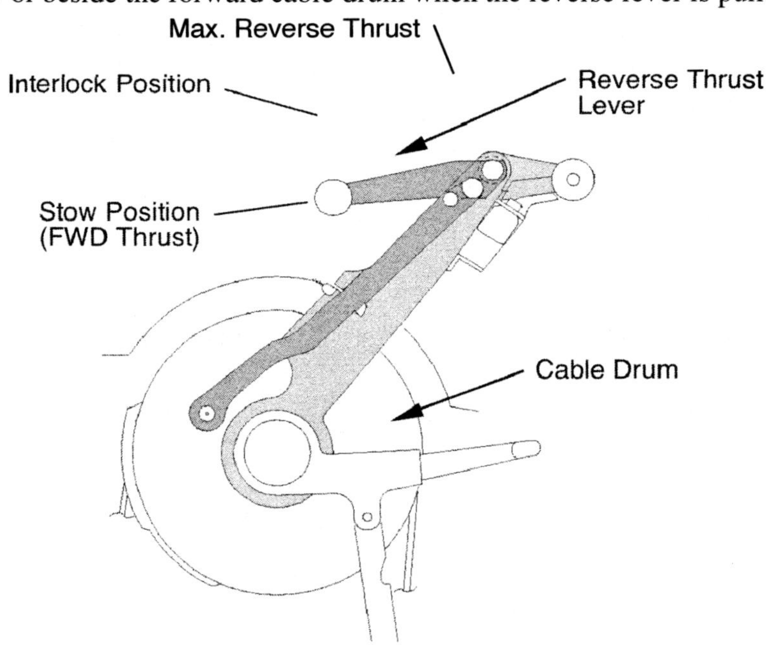

Fig. 6.15 The main positions of the reverse thrust lever of a B737

from the stow position. The reverser control switches are connected to the reverser control circuit in the aircraft. This circuit controls the deploy and the stow sequence of the reverser.

The movement of the reverse lever leads to a travel of the cable system in the opposite direction compared to the forward thrust selection. The travel area of the reverse thrust lever is divided into two sectors. For the selection of reverse thrust the lever can only be pulled up to the interlock position. In this position the rear drum touches a movable interlock stop until the reverser is fully deployed. The input to the FCU at this position is, despite the lever movement, idle power. The reverser has a mechanical connection to the movable interlock stop. When the reverser is fully deployed, the interlock stop has moved and the cable drum is free to turn further into the reverse direction. The pilot can select the required power setting by moving the reverse thrust lever through the second sector of the lever travel area. The interlock position stop prevents the selection of a higher than idle engine power until the reverser is fully deployed.

6.2.3 Electrical Thrust Lever Systems

6.2.3.1 Forward Thrust Control

In thrust lever systems for FADEC engines the thrust lever angle is sensed electrically by the EEC. Basically the main components of the system for each engine are the thrust lever and the twin thrust lever angle resolver. The latter is mechanically connected to the thrust lever and its harness connects the resolver to the EEC channels A and B. Between the thrust lever and the resolver an artificial feel unit is installed to provide the required friction. All these mechanical components of the system are installed between the thrust levers and the cockpit floor. The power for the resolver excitation is supplied by the EEC. Thus the thrust lever angle resolver is operable as long as the EEC is supplied with electrical power.

In their system designs Boeing and Airbus made different approaches concerning the automatic thrust control and the reverser interlock. The Boeing design resembles the traditional mechanical thrust lever system for the pilot's view. The mechanics of the system contain an autothrottle servo and a reverser interlock. The input for the power setting to the EEC during autothrottle operation comes from the autothrottle servo through the thrust lever movement. This principle requires the necessary mechanical components in the thrust lever system.

In Airbus aircraft the thrust lever systems are designed not to move during the operation with automatic thrust control. Therefore the automatic thrust control mode in Airbus aircraft is called *autothrust* and not *autothrottle*. The thrust levers can only be moved manually. To find the lever positions used after take-off quickly and easily, detents are located in the artificial feel unit at the maximum continuous position and at the climb thrust position. In Fig.6.16 the thrust lever systems of both companies are shown.

6.2.3.2 Reverse Thrust Control

When reverse thrust is selected with an electrical thrust lever system, the resolver is turned from idle in the rearward direction towards lower thrust lever angle values. The reverser operation can be controlled by the EEC or by the aircraft mounted reverser control system. In Boeing aircraft the latter is used and the control switches described earlier are installed at the thrust levers. In Boeing aircraft the mechanical interlock for the reverser control is also installed. This function is realized by one interlock solenoid or interlock actuator for each reverse thrust lever. These devices are controlled by the related EEC.

In Airbus aircraft the EEC controls the reverser operation. Here only the thrust lever angle resolver is required for the selection of reverse thrust.

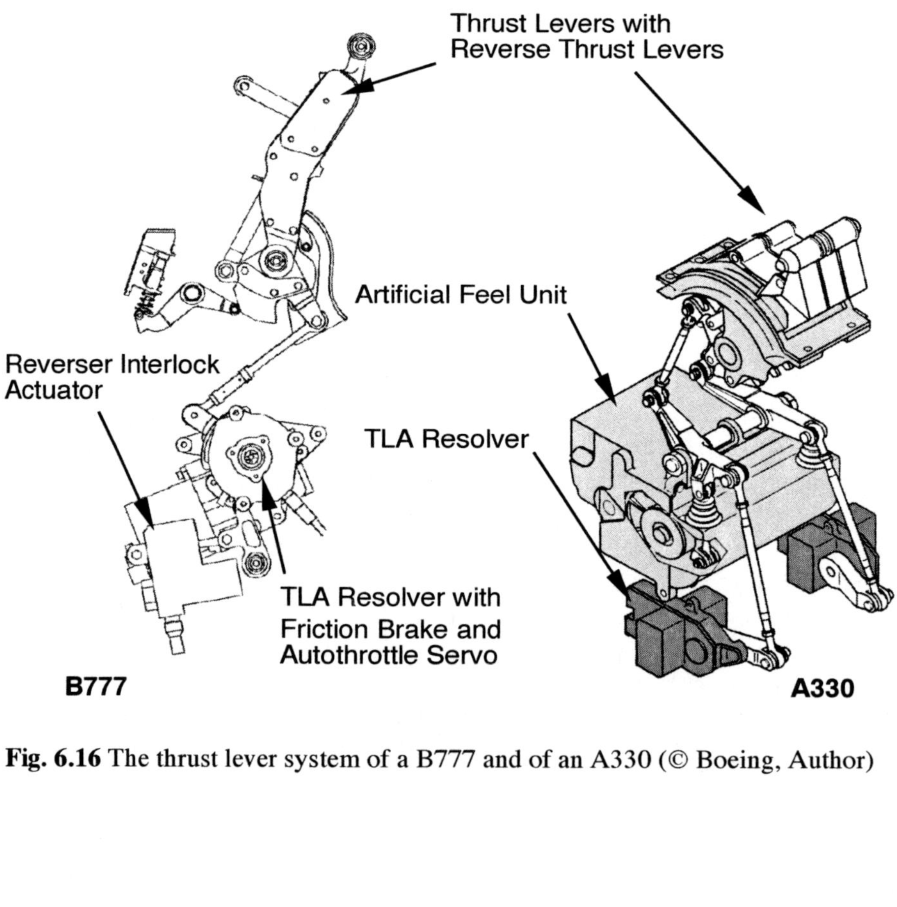

Fig. 6.16 The thrust lever system of a B777 and of an A330 (© Boeing, Author)

The thrust lever angle is sensed additionally by other computers to secure the system against accidental actuation. The function of the interlock is established in the software. There is no need for a mechanical interlock in the thrust lever system. After the reverser is fully deployed the EEC sets the engine rotor speed according to the resolver angle. Because no interlock is installed, the thrust lever can be moved directly to the appropriate position for the required power setting.

In aircraft of the A320 family the thrust lever system is simplified one step further. On the thrust levers of the A320 the reverse thrust levers are eliminated. For the selection of reverse thrust the thrust levers are moved behind the idle stop into the reverse range. To prevent an accidental movement of the levers into the reverse range, the idle stops are installed. For the selection of reverse thrust the idle stop is released by the use of the latching lever on each thrust lever. Figure 6.17 shows these thrust levers in the climb thrust position for autothrust operation during flight.

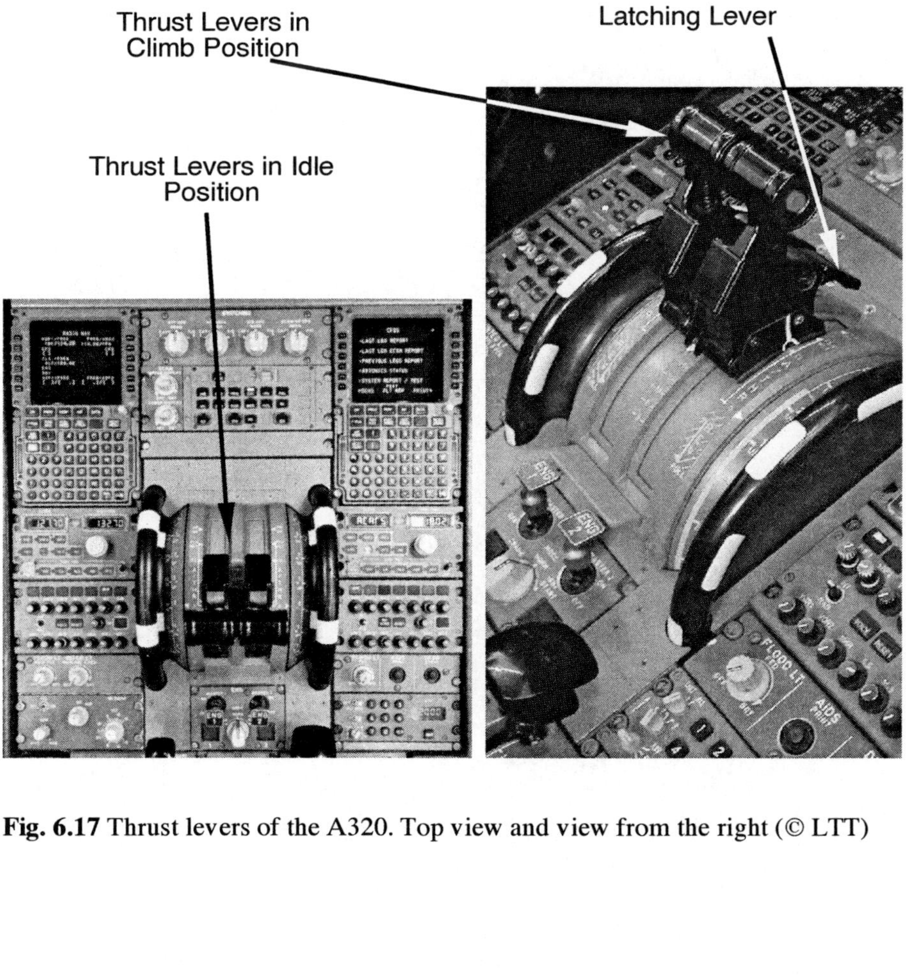

Fig. 6.17 Thrust levers of the A320. Top view and view from the right (© LTT)

6.2.4 Control Switches

Additionally to the thrust levers and the HP shut-off valve controls the following switches related to the engine controls are installed in the cockpit:

- Take-off/go-around switches
- Autothrust disconnect switches
- Start switches
- Start/ignition mode selector (Airbus aircraft)
- EEC mode switches
- FADEC ground power switches

One take-off/go-around switch is installed on each thrust lever in Boeing aircraft. If one of the switches is pressed on ground, the autothrottle system accelerates all engines to take-off thrust. Simultaneously the flight director changes into the take-off mode. If one switch is pressed during approach, the autothrottle system accelerates all engines to go-around thrust and the flight director changes to the go-around mode. In Airbus aircraft both flight maneuvers are initiated by moving the thrust levers manually into the take-off position.

The manual control of the autothrottle or autothrust systems is possible with the appropriate switch on the auto flight panel. If the pilot wants to take over the thrust control by a quick grip on the thrust levers, he can disconnect the autothrottle (or autothrust system) by pressing one disconnect

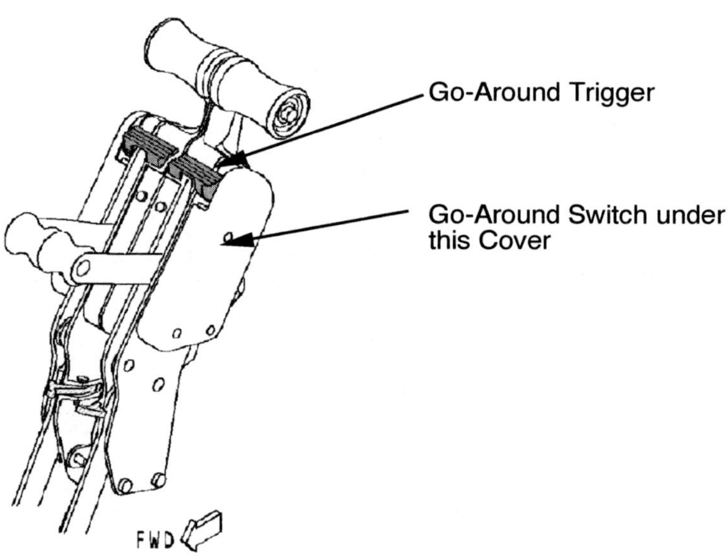

Fig. 6.18 Go around switches in the thrust levers of a B777 (© Boeing)

button on a thrust lever knob. For this purpose a disconnect switch is installed in the thrust lever knobs of engine No.1 and No. 2 (or No.4 if the aircraft has 4 engines).

The start switches and other controls used for engine start like the mode selector in Airbus aircraft, can also be included in the group of engine controls. Their functions are described in Chapter 7.

For some FADEC engines EEC mode switches are required in the cockpit to force the EEC manually into the alternate operating mode. If such switches are installed in the aircraft depends on the design of the EEC.

They are used if one EEC has changed into the alternate mode due to a system fault. In this case the other EECs are also forced into the alternate mode to operate all EECs in the same mode.

While the aircraft is parked on the ground the power supply for the EECs is switched off. For retrieval of fault messages stored in the EEC and FADEC tests during troubleshooting the EEC needs electrical power. This power supply can be switched on independently from the other engine controls with the ground power switch for each EEC.

6.2.5 Thrust Control during Automatic Flight

6.2.5.1 The Flight Management Computer

The aircraft may be controlled by the crew itself by using the controls or the crew uses the automatic flight system for aircraft control. In the latter case the crew may use the multipurpose control and display units (MCDU) as the principal interface with the aircraft and the auto flight system. The flight management computer (FMC) is the computer the crew communicates with via the MCDUs. It interfaces with the auto-thrust (or auto-throttle) system and the autopilot system. The FMC isolates the crew from the control loop by assigning a managerial and supervisory role to it within an integrated automatic flight management system. This automated operation provides greater precision of flight path and engine control.

A flight management system integrates the functions of navigation, performance management, flight planning and three-dimensional guidance and control along a pre-planned flight path.

Besides a large navigation data base the system uses a detailed database of aerodynamic and engine data of the specific aircraft. This data base contains flight profile data for such flight phases as take-off, flex-take-off, climb, cruise, descent, holding and go-around. The system takes also into account non-standard temperatures and bleed air demands. For the cruise mode specific data for maximum economy, long range and selected speed /

Mach number (if mandated by air traffic control) are available. The crew prescribes a cost index, which is a trade-off between fuel cost and flight time related cost. Models for the wind and temperature profiles along the flight path are used, which in the planning phase are based on data inserted by the crew. Later on during the flight the system uses actual values measured with the installed sensors.

Thus the system is capable of planning ahead through the entire flight generating a flight plan that minimizes costs. During the flight alterations to the active flight plan, based on in-flight data, are proposed with the aim to minimize trip costs. Through the MCDU the pilots monitor the progress along the flight path, obtain messages and suggestions (e.g. for a step climb) and receive predictive information (e.g. fuel remaining on arrival). They can also investigate the consequences of alternatives of the active flight plan.

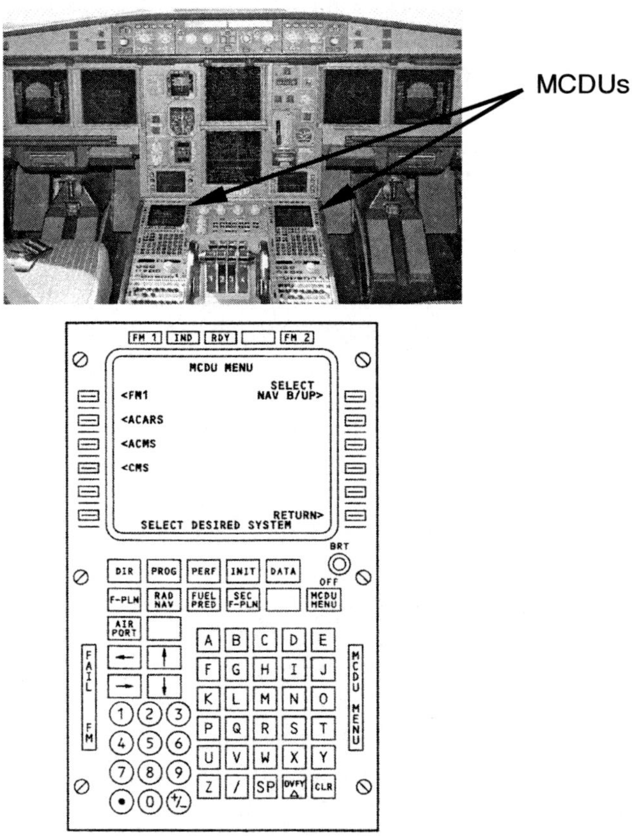

Fig. 6.19 MCDU of an A340-600 for the interface with the flight management computer (© LTT)

When activated by the flight crew the flight management system sends steering commands, affecting flight plan execution, to the autopilot and the autothrottle system. The engine thrust setting was calculated precisely by the FMC for the respective flight phase before the auto-throttle system is commanded. In cruise flight minor corrections to the speed / Mach number are not made by a thrust command but by an elevator deflection commanded by the autopilot. The airplane is allowed to gently climb or descend within a ±50 ft band to correct minor airspeed deviations. The elevator is much more sensitive to small changes than the engine. Therefore the thrust setting remains unchanged for such a small airspeed correction. Thus the engagement of the flight management system ensures the most economic engine operation.

6.2.5.2 Automatic Thrust Control with a Mechanical Thrust Lever System

For the automatic selection and adjustment of the thrust setting for a flight maneuver the *autothrottle system* is installed in the aircraft. In mechanical thrust lever systems the autothrottle system mainly consists of the autothrottle servos that are controlled by the flight management computer or a separate autothrottle computer. The autothrottle servos are installed in the cable system of the thrust lever systems. The servo motors are not permanently engaged to the cable systems. They engage with a clutch when the autothrottle system is activated by the pilot. Then the autothrottle computer selects the power setting demanded by the flight management computer. During the operation of the autothrottle system the pilot can override the thrust lever position directly by moving the thrust lever manually. To allow this direct override by the pilot, a friction clutch is installed between the servomotors and the cable drums of the autothrottle servos.

6.2.5.3 Automatic Thrust Control with Electrical Thrust Lever Systems

In Boeing aircraft like B747 or B777 the autothrottle system operates basically the same way as in aircraft with non-FADEC engines. The flight management computer controls the thrust lever angle with the autothrottle servo and the EEC controls the engine to the commanded thrust with regard to the thrust lever angle. The autothrottle servo with its clutch is installed beside the thrust lever angle resolver. It moves the whole thrust lever system during operation.

In Airbus aircraft the input for the power setting (EPRcommand or N1command) is sent from the auto flight system directly to the EEC via the data bus connection. During the autothrust operation the thrust levers remain in the last set position. The autothrust operation is activated after take-off when the thrust levers are moved back to the max. climb thrust or the max. continuous thrust position. With the thrust lever in the max. climb thrust position, which is the normal position after take-off, every thrust setting between max. climb and idle is available during autothrust operation.

Another function in Airbus aircraft, that influences the thrust setting, is the Alpha-Floor Protection of the flight control system. The flight control system ensures a flight envelope protection during all flight phases. When the airspeed reaches the lower limit the flight control system commands the engine thrust to maximum take-off, despite the thrust control mode of the engine, to accelerate the aircraft. This function overrides all other thrust setting inputs, e.g. it even operates when the thrust is set manually.

6.2.5.4 Reduced Take-Off Thrust

With all engines operating, an aircraft has more installed thrust than necessary for a take-off with low weight or from a long runway. This fact is the basis for lowering the operating costs of an engine by using less than the maximum take-off and climb thrust. The operation with lower take-off thrust settings reduces wear on the engine and extends its life. Thus the maintenance cost as part of the operating costs of an engine are reduced. During normal airline operation a certain percentage of the take-offs are done with reduced take-off thrust.

There are basically two methods to operate with reduced take-off thrust. One method is the *flexible takeoff thrust* method (or assumed temperature method as it is called by Boeing). But this method cannot be used on contaminated runways.

To further optimize the thrust for takeoff on short or contaminated runways and during climb, the features permitting to automatically reduce the thrust during takeoff and climb are the *derated takeoff thrust* and the *derated climb thrust*

Flexible take-off thrust is a thrust level less than the maximum take-off thrust for which the takeoff performance is established by approved simple methods such as adjustments or corrections to the maximum takeoff performance.

For aircraft operations it is important that the thrust setting parameter used with the flexible take-off method for establishing thrust for take-off, is not considered as take-off operating limit.

Before the reduced thrust with the flexible take-off method can be used, the pilot must enter the *flexible take-off temperature* into the flight management computer. This is done on the take-off performance page of the

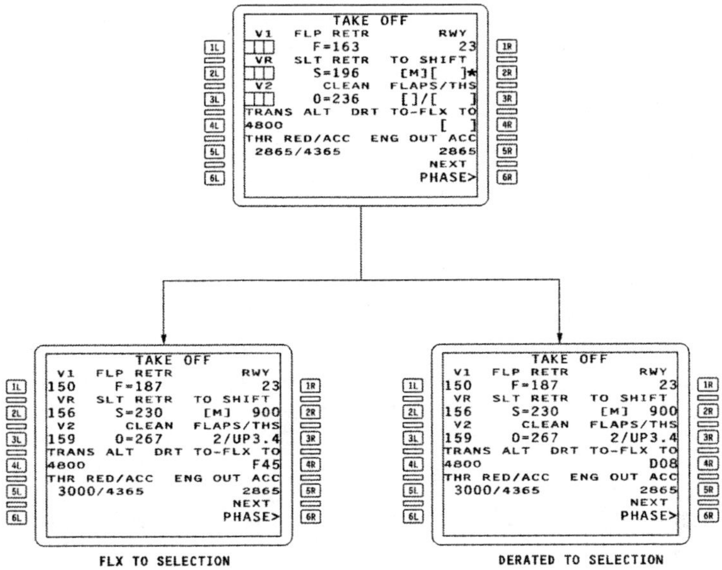

Fig. 6.20 The take-off performance page of an A340-600 for the selection of reduced take-off thrust

FMC indicated on the MCDU. Figure 6.20 shows this page on the MCDU of an A340. The flexible take-off temperature must always be a temperature higher than the flat rate temperature of the engine.

The result for the engine control system during take-off will be, that it assumes the flexible take-off temperature as the actual ambient air temperature. Thus the control system controls the engine to a thrust belonging to the flexible take-off temperature on the flat rate curve of the engine. Figure 6.21 shows the flat rate curve with the flexible take-off temperature.

The derated take-off thrust is a thrust level less than the maximum take-off thrust for which there is in the airplane flight manual a set of separate and independent, or clearly distinguishable, take-off limitations and performance data, which comply with all the takeoff requirements of the certification standards.

When the reduced take-off thrust is set according to the derated take-off thrust method, the value of the thrust setting parameter, which establishes thrust for take-off, is presented in the flight manual. The established thrust setting is considered as a new operating limit for take-off.

For the preparation of the aircraft to use the derated take-off

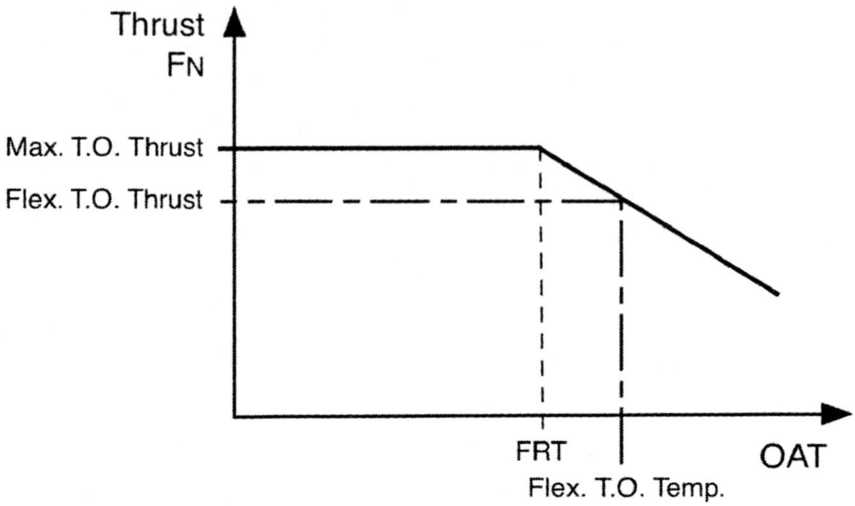

Fig. 6.21 Thrust/OAT curve of a flat-rated engine with flexible take-off temperature

thrust method the pilot must enter the derate level into the FMC. On Airbus aircraft this is done by entering the letter D with the derate level in percent (D08 for example) into the FMC.

7 Starting and Ignition

To make the start of an engine possible, two systems are installed for this purpose on a turbine engine. These are the start system and the ignition system. These two systems are always used together for an engine start on ground, but in the typical system description in an aircraft maintenance manual they are described in different chapters.

7.1 The Starting System

7.1.1 General

The starting systems on the large turbofan engines are mainly pneumatic starting systems. This type of system has the advantage of a very low weight of the starter and the other system components compared to an electrical start system. The engines of the Boeing 787 have no bleed air system to prevent the decrease in engine efficiency during bleed air use. A bleed air extraction reduces the thermal efficiency of an engine. Thus there is no traditional pneumatic system connected to the engines. All the energy the engines deliver for the aircraft systems is electric energy. The engines are started with the two generators on each engine. They are designed as starter generators. The large generators are necessary for the supply of the aircraft and are no additional weight for the engine start system.

On most engines the accessory gearbox is driven by the HP spool via a drive shaft. Thus a starter motor (or in short *starter*) mounted on the accessory gearbox is able to drive the HP spool. The start sequence of an engine, as shown in Fig. 7.1, begins with the acceleration of the HP spool to a speed high enough to generate an airflow through the combustion chamber that allows the build-up of an inflammable fuel/air mixture. This is safely possible at relatively low rotor speeds of approximately 20%.

After the shaft speed for safe light-off is reached, the high pressure shutoff valve is opened and the fuel flows to the fuel nozzles. Some percent of HP spool speed earlier or simultaneously, the ignition is activated and light-off occurs as the fuel reaches the combustion chamber. At this low

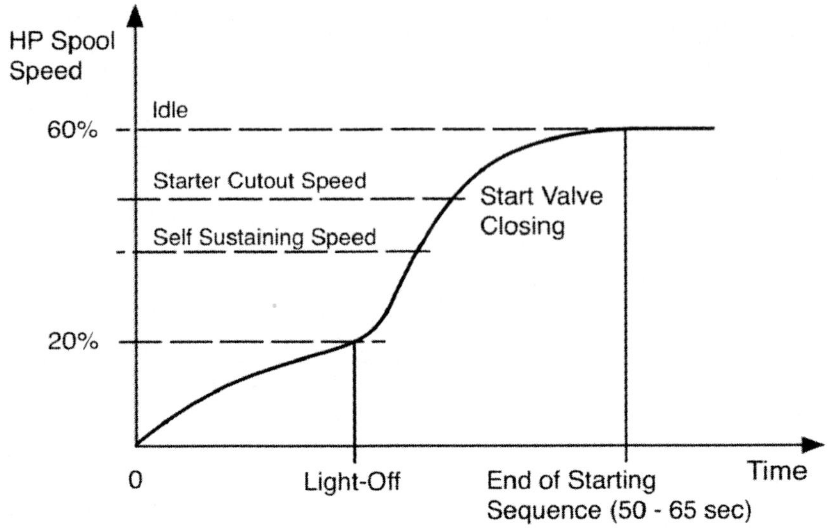

Fig. 7.1 Acceleration of the high pressure spool during the start sequence. The typical speeds that are important during starting are shown

rotor speed the torque of the HPT is too low for the further acceleration of the spool. Thus the starter remains on after the combustion has begun. The acceleration with the combined torque of the starter and the HPT continues to a rotor speed from which the engine can accelerate without any further starter assistance. This *starter cutout speed* lies above the *self sustaining speed* of the engine. At the starter cutout speed the starter and the ignition are switched off by the EEC and the engine accelerates to idle.

After the beginning of the combustion the EGT starts to rise. It reaches its peak before the HP spool speed reaches the starter cutout speed. For a trouble-free start sequence it is important to release the fuel flow at or above the rotor speed intended for this step. Otherwise there is a risk of a compressor stall with high EGT. Even without a compressor stall the EGT limit can be exceeded, if the airflow and thus the torque of the HPT is too low.

7.1.2 Starting System Components

The typical pneumatic starting system has the following components:

- Starter duct to connect the starter with the aircraft pneumatic system
- Starter
- Start valve for the shut-off of the air supply with the associated wiring
- Cockpit controls and indications

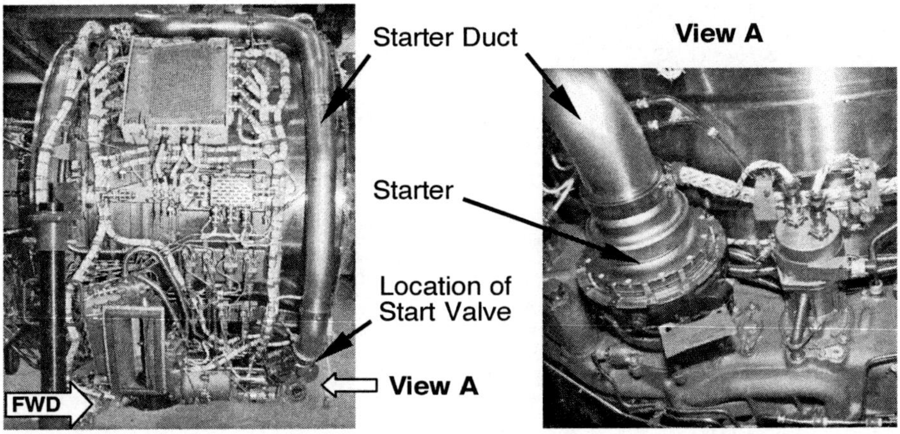

Fig. 7.2 The starter duct and the starter on a V2500-A5. The starter is mounted on the forward face of the accessory gearbox (© LTT)

The system components installed on the engine are often part of the power plant build-up kit, which is necessary for the adaptation of the engine to the aircraft.

7.1.2.1 The Starter Duct

The starter duct connects the starter with the pneumatic system of the aircraft. Depending on the size of the engine and the location of the start valve, the starter duct consists of two or more segments as shown in Fig. 7.2. It is mounted to the engine case with some links. The upper segment is installed after the engine is fitted to the engine pylon of the aircraft.

7.1.2.2 The Pneumatic Starter

The pneumatic starter contains a turbine, a reduction gear and a clutch that connects the reduction gear with the starter output shaft during starter operation. The starter output shaft fits into the associated gear shaft of the gearbox. To ensure a quick replacement of a damaged starter, it is mostly attached with a mounting V-clamp to the gearbox adapter. Figure 7.3 shows this type of attachment.

After the shut-off of the starter air supply during the start sequence the clutch operates as a ratchet freewheel. This prevents the powering of the turbine by the engine. During further acceleration of the engine the claws are opened by centrifugal force and the starter gear is completely disconnected from the gearbox shaft during engine operation. Fig. 7.4 shows the arrangement of the reduction gear and the starter clutch.

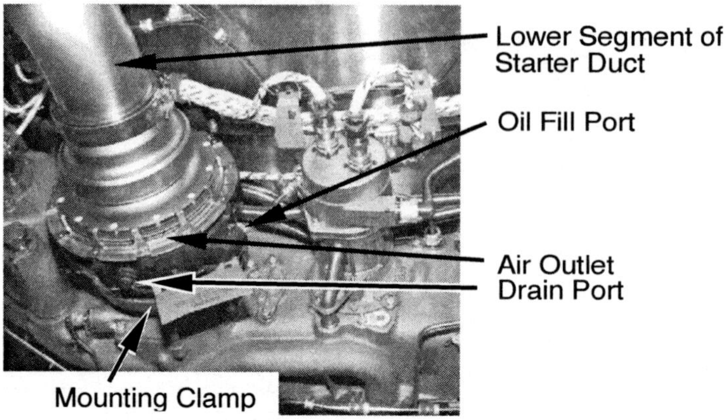

Fig. 7.3 The starter on the accessory gearbox of a V2500-A5 (© LTT)

For lubrication the starter contains a small amount of engine oil. This small oil filling heats up very quickly. Thus the starter operating time for continuous operation is limited. Usually three consecutive start cycles with two minutes pausing between them are permitted. These cycles must be followed by a cool down time of 30 minutes before the next start attempt can be made.

The typical pneumatic starter has no electrical components. Thus it has no electrical connections.

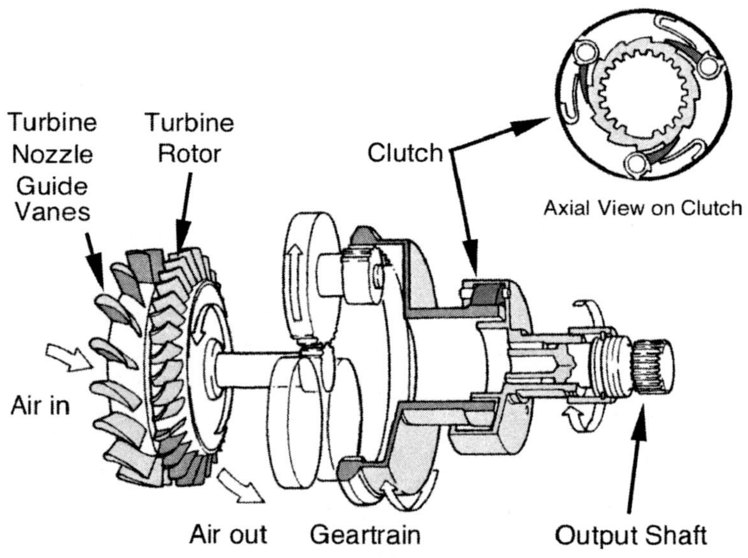

Fig. 7.4 The main components of a pneumatic starter. The design of the clutch and its location within the starter are shown

7.1.2.3 The Start Valve

The start valve is a pneumatically operated butterfly valve. It is commanded electrically via its solenoid valve that controls the supply of air pressure to the valve actuator. The valve operates with the air pressure of the starter duct that is also known as *duct pressure*.

For the indication of the valve position it contains switches that are connected to the start valve open light in the cockpit or to the EEC on most FADEC engines.

The start valve is installed in the starter duct upstream of the starter. Where in the duct the valve is installed, depends on the reachability of the valve while the aircraft is on the ground. The valve must be reachable by the mechanic on the ground for the manual operation of the valve during starting if the electrical operation is not possible. On aircraft with the engines low above the ground the start valve is installed near the horizontal centerline of the engine. On large aircraft the start valve is installed at a lower position, because the valve can only be reached from below. All starter valves have provisions for the manual operation of the valve like handles or tool adapters. In Fig. 7.5 the start valve installation of a V2500 is shown. The location of the valve actuator and the manual drive square is visible.

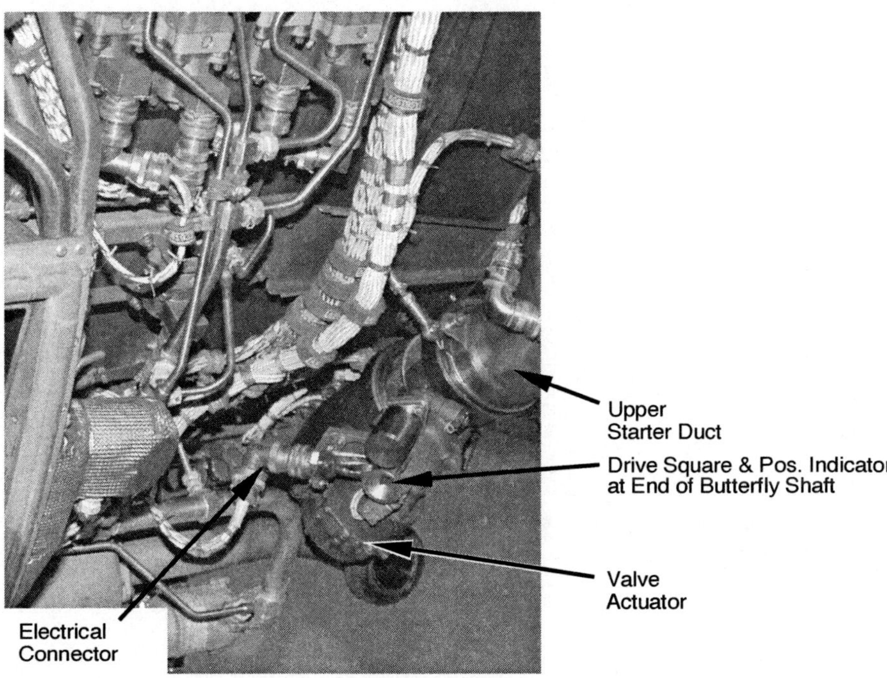

Fig. 7.5 The pneumatic start valve of a V2500-A5 (© LTT)

7.2 The Ignition System

7.2.1 General

The ignition system is used to ignite the fuel/air mixture in the combustion chamber. The ignition system is engaged in three situations for:

- Engine start on ground
- In-flight start after a flameout
- Continuous ignition operation

Continuous Ignition is used under safety-critical conditions (take-off, landing and adverse weather). The most demanding and critical situation is the relight at high altitude. The low temperatures encountered cause a decrease in volatility, which makes it difficult to ignite the fuel. To ensure relight even in such a situation requires a powerful and reliable ignition system.

The ignition of the fuel/air mixture is accomplished with sparks emitted by an igniter plug installed in the combustion chamber wall. The high voltage supply for the igniter plug comes from an ignition exciter installed on the engine. Two identical ignition systems are installed on an engine. Figure 7.6 shows the ignition system components on a V2500-A5.

7.2.2 Ignition System Components

Each ignition system on an engine consists of the following components:

- Ignition exciter with electrical power sapply
- Ignition lead (high tension cable)
- Igniter plug

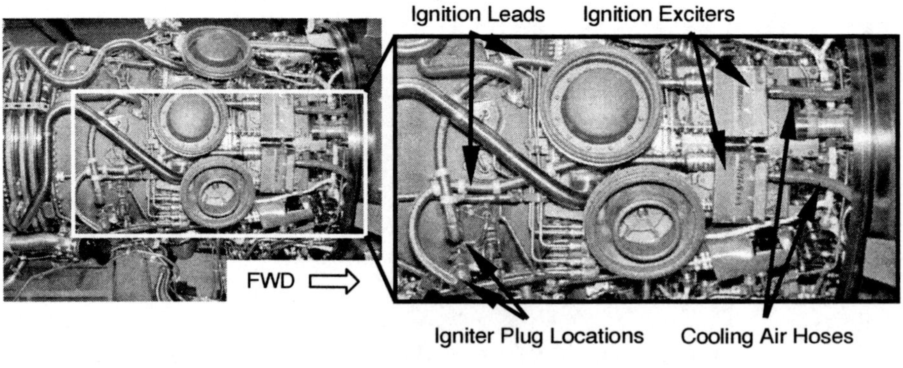

Fig. 7.6 The ignition system components on a V2500-A5 (© LTT)

7.2.2.1 The Ignition Exciter

The ignition exciters are installed on the engine fan case or on the core engine. If the ignition exciters are installed on the fan case, the air temperatures around the exciters are lower and no exciter cooling is necessary. When this installation location is used, the ignition leads are longer compared to the core engine installation. These long ignition leads are usually cooled in the core engine area only. This type of installation is used on the CFM56-7B. The ignition exciters are shown in Fig. 7.7.

The installation on the core engine results in very short ignition leads. To have the high-tension cables as short as possible minimizes the energy loss between exciter and igniter plug and also the chance for shorts to ground due to a damaged cable. But exciters installed on the HPC case, like these of the V2500 shown in Fig. 7.6, operate in a very warm environment and must be air cooled. To guide the cooling air around the exciter housing a cooling shroud is installed around the exciter. They are usually cooled with fan air that exits the exciter cooling shroud into the ignition lead.

The ignition exciters are supplied with 115 VAC from the aircraft. They contain capacitors, the primary and secondary coil and a spark gap. The charging of the capacitors lasts approximately 1 second. Then they discharge across the spark gap to the igniter plug and the charging starts again. Due to the charge of the capacitors this discharge releases a high level of energy. The current level to the igniter plug is around 1500 A. Thus it is strictly prohibited to work on a powered ignition system. This would be extremely dangerous.

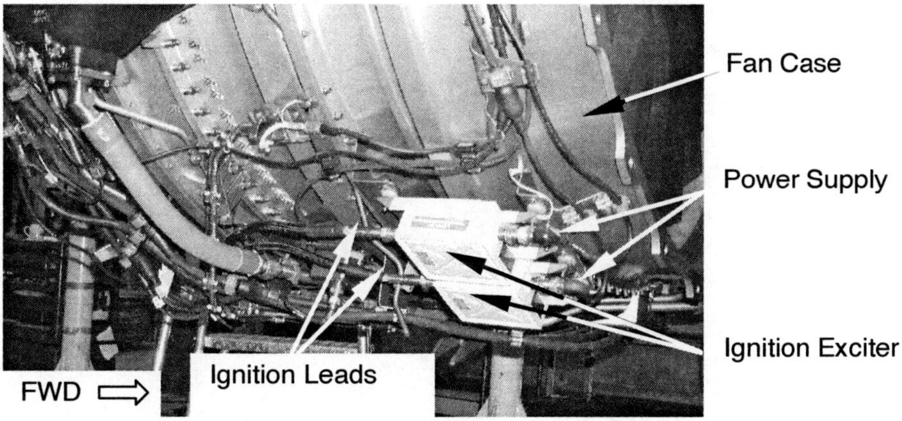

Fig. 7.7 The ignition exciters on the fan case of a CFM56-7B (© LTT)

The data of a typical ignition exciter:

- Energy rating 2 to 20 Joule
- Output voltage 2000 V
- Output current 1500 A
- Spark rate 1 per second
- Power supply 115 VAC

7.2.2.2 The Ignition Lead

The ignition lead conducts a very high current from the ignition exciter to the plug over a distance of up to several meters. All electrical connections must always be in good condition to keep the resistance at the connectors low. Lost energy results in sparks with decreased energy. The ignition lead has an internal copper wire that is connected to the center contacts in the connectors at both ends. On the outside the lead has a flexible steel conduit as a mechanical protection. It also operates as an electromagnetic shielding. Between the flexible conduit and the internal copper wire a silicon rubber insulation embeds the center wire. When the plug connector is attached to the plug with the coupling nut, the center contact touches the contact cap of the plug to close an electrical connection. A ceramic insulator keeps the contact centered.

Ignition leads are cooled where they can become very warm. The cooling of the lead is necessary because the copper wire increases its resistance

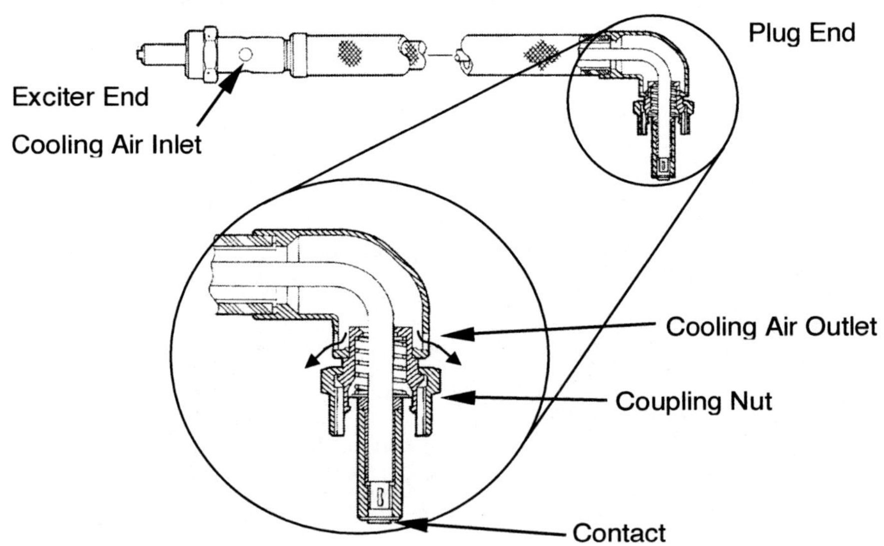

Fig. 7.8 A typical ignition lead

with increasing temperature. At the cooled portion of a cable air from the LPC exit or from the fan duct passes through the gap between the conduit and the rubber insulation. The short cables of core engine mounted exciters, as shown in Fig. 7.8, are cooled over the full length. The longer cables of fan case mounted exciters are air cooled along the core engine only. The cooling air for such an ignition lead enters through inlet holes and exits the lead above the coupling nut. From here the air flows over the surface of the igniter plug and cools the outer surface of the plug.

7.2.2.3 The Igniter Plug

For the location of the two igniter plugs of an engine two different configurations are in use. They are installed on one side of the combustion chamber or symmetrically left and right of the engine centerline. In both configurations the igniter plugs are installed below the horizontal centerline of the engine in the combustion case. The plugs are screwed into a boss in the combustion case and its inner end immerses a few millimeters into the combustion chamber. This immersion is limited because otherwise the inner end of the plug would be too hot during engine operation with the result of a short lifetime of the igniter plug. Figure 7.9 shows the location of an igniter plug in the combustion chamber of a V2500-A5. To achieve a sufficient cooling of the igniter plug tip, air from the combustion chamber secondary flow enters the plug and flows through the tip into the combustion chamber.

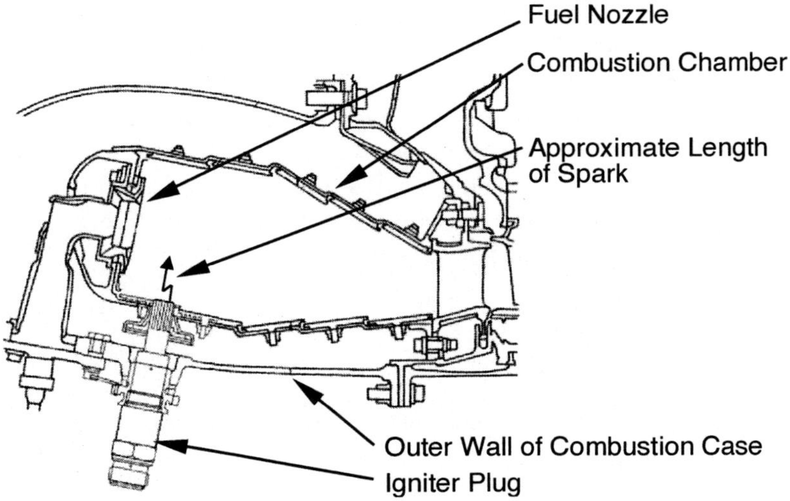

Fig. 7.9 Location of the igniter plug in the combustion chamber of a V2500-A5 (© LTT)

Due to this limited immersion depth into the combustion chamber, the igniter plug itself is not located within the fuel/air mixture. To ignite the fuel/air mixture the spark shoots out of the plug into the combustion chamber. This is possible due to the high amount of energy stored in the capacitors of the exciter and the design of the igniter plug tip with the two electrodes. Here between the two electrodes a semiconductor layer covers the isolator. This layer facilitates the ionization of the air that lowers the required energy level for the sparking.

7.2.2.4 Power Supply

The power supply of 115VAC for the ignition exciters comes from the electrical network of the aircraft. On non-FADEC aircraft the current for the exciter passes the start switch and the start lever switch before it reaches the exciter. On FADEC aircraft the current passes the disconnect relay in the interface computer and then the EEC relay. Thus the ignition can be activated and deactivated by the EEC during engine start and during engine operation.

7.3 Cockpit Controls and Indications

The start system is controlled by using a start switch on the overhead panel or by the mode selector and engine master switch used on Airbus aircraft. On non-FADEC aircraft the start switch also activates the power supply to the ignition systems of the engine. On FADEC engines the power supply for the ignition exciter is usually switched by the EEC.

For a better understanding of the operation of the start and the ignition system three examples of these systems are described in the following. The systems of three different engines are used as examples. In these system descriptions the use of the system controls is explained, too. The systems shown below are used on these engines:

- CFM56-3 on the B737
- CFM56-7B on the 737 NG
- V2500-A5 on the A320

Each description begins at the point when the preparations for the engine start are finished, e.g. the electric power supply and the air supply is established. The system schematics show the signal flow for the engine No.1 of the respective aircraft.

7.3.1 Starting System and Ignition System of the CFM56-3

7.3.1.1 Ground Start

The start sequence begins with the opening of the start valve for the activation of the starter. For the opening of the start valve the start switch is set to the GRD position. This switches the 28V to the start valve of the respective engine and the start valve opens. The open start valve position is indicated in the cockpit by the start valve open light. The start switch remains in the position GRD until the N2 indicator releases the solenoid of the start switch at the starter cutout speed (see Fig. 7.10 for system details).

With the start switch in the GRD position the power supply for the two exciters can pass the start switch up to the start lever switches. If one or both exciters are energized depends on the position of the selector switch between the two start switches. Normally one igniter plug only is used for a ground start. What is also often done is to alter the igniter for the ground start from one start to the next with the selector switch. With this procedure the wear on both igniter plugs is identical and a faulty system can be identified on the ground during normal operation.

After the engine has reached the N2 speed for light-off, the pilot moves the start lever to the run (or idle) position. This releases the electrical power to the exciters and opens the high pressure shut-off valve within the MEC.

The exciters are powered until the start switch moves back to OFF at the starter cutout speed. This happens at 43% N2 when the N2 indicator deenergizes the holding solenoid of the start switch. This also causes the start valve to close.

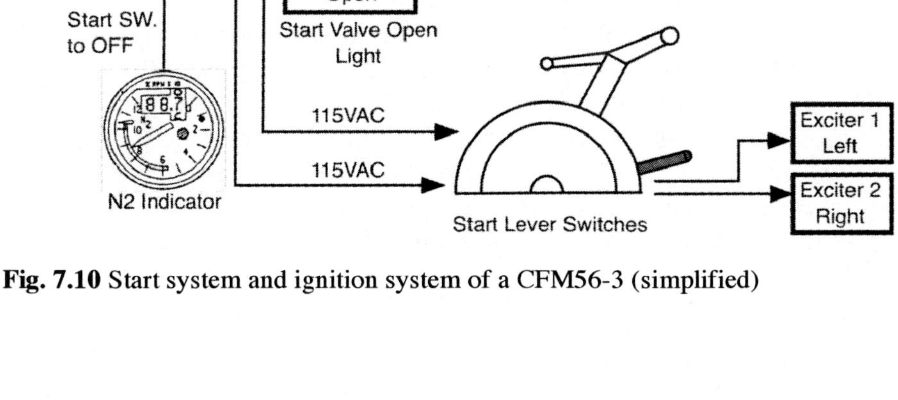

Fig. 7.10 Start system and ignition system of a CFM56-3 (simplified)

7.3.1.2 In-Flight Start

For an in-flight restart of the engine the pilots have the choice between a starter assisted engine start and a non-starter assisted engine start. The variant without starter assistance is used, when the windmilling speed of the engine is high enough for a light-off. In this case the start switch is positioned to the position FLT. Otherwise the position GRD is used. In the position FLT the selector switch has no function. Both exciters are always energized when the start switch is in this position.

7.3.1.3 Continuous Ignition

Normally the ignition system is not used during most of the operating time of the engine to extend the life time of the igniter plugs. But to prevent a flameout in critical flight phases and in severe weather the ignition system is switched on to operate in the mode continuous ignition. For this purpose the start switch is turned to the position CONT.

7.3.2 Start System and Ignition System of the CFM56-7B

7.3.2.1 Ground Start

The controls to operate the ignition and the start system on the B737NG are the same as for the CFM56-3 on the older B737. The functions of the start switches have changed a bit. The power for the start valve solenoids is the only electrical power switched directly by the start switches. The power supply for the ignition exciters is routed via the start lever switches and the ignition relays within the EECs.

The position of the start switch and of the selector switch is sensed by the DEUs. These computers send this information to the EEC as digital data. The EEC also receives the start switch position information directly from the switch. The EEC uses the information direct from the switch alternately if the DEU data cannot be received due to a system fault. Figure 7.11 shows the simplified system schematic.

When the start switch is turned to the ground position the start valve opens and the EEC is powered from the aircraft network. At the time the light-off N2 is reached the EEC is fully operable. At this speed the start lever is moved to the idle position by the pilot to release the fuel flow. Simultaneously the power supply for the exciter passes the start lever switches and reaches the ignition relays within the EEC. The EEC has received the start lever position information and closes the ignition relays.

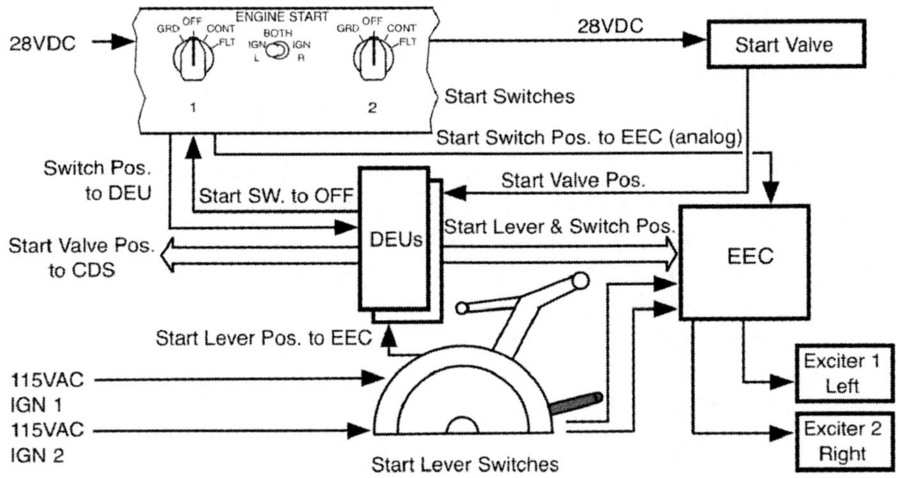

Fig. 7.11 The start and ignition system of the CFM56-7B (simplified)

Thus one or both ignition exciters are powered and the engine lights off. At the starter cutout speed of 55% N2 the DEUs command the start switch return to OFF. The engine accelerates to the idle N2 and the start sequence ends.

7.3.2.2 In-Flight Start

The switching for the in-flight start is the same as described for the CFM56-3. Both exciters are always energized during an in-flight start. The pilots must also decide if a windmilling start is possible or a starter assisted start is necessary. If they select the windmilling start with the start switch in the FLT position and the windmilling N2 of the engine is too low, the EEC sends a message to the display that a cross bleed start with the starter is required.

7.3.2.3 Continuous Ignition

For the manual selection of continuous ignition the same procedure as shown at the CFM56-3 is used. In aircraft with an automatic ignition feature the EEC turns on the ignition for take-off and landing. It receives the related signal from the airframe systems. On such aircraft the OFF position of the start switch is labeled AUTO.

If the EEC detects a flameout during the engine operation it activates the ignition system by closing both ignition relays.

7.3.2.4 Limit Protection by the EEC

A very useful feature on a FADEC engine is the limit protection during starting provided by the FADEC system. During the ground start sequence the EEC is able to detect a *hot start* and a *wet start*.
If the EEC detects the symptoms of a hot start, it shuts off the fuel flow and prevents the exceedance of the EGT limit. A wet start is detected by the EEC if the EGT rise is not sensed after a specific time after fuel flow release. In this case the EEC closes the high pressure shut-off valve and continues the motoring of the engine to blow the liquid fuel out of the combustion chamber. This blow-out prevents a hot start during the next engine start.

The only limit protection during a flight start is the indication of the EGT limit exceedance.

7.3.3 Start System and Ignition System of the V2500-A5

The controls for the starting system and the ignition system are different in Airbus aircraft compared to the Boeing design described before. The integration of the FADEC system into the airframe systems is also different as shown in Chapter 5. Figure 7.12 shows the simplified schematic of the V2500-A5 starting and ignition system.

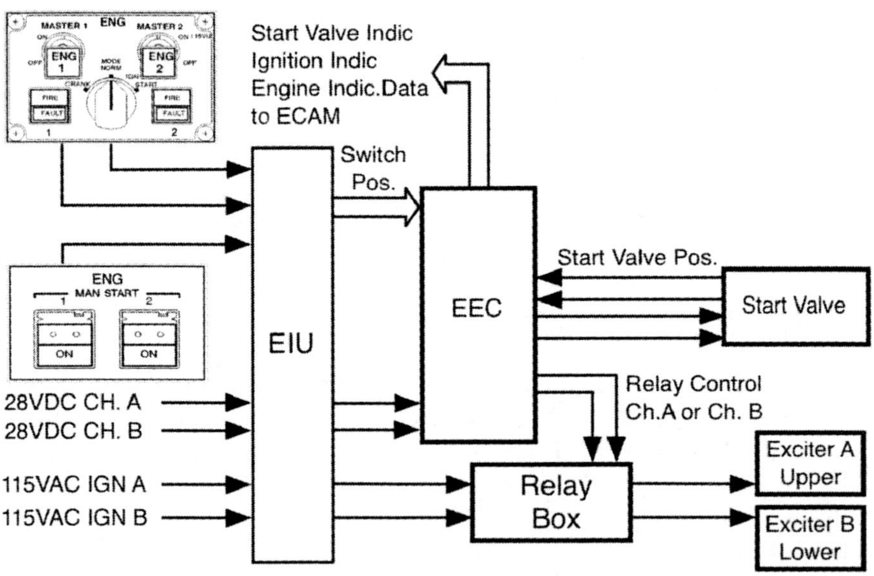

Fig. 7.12 The start and ignition system of the V2500-A5 (simplified)

For the start of an engine three control elements are provided. These are the engine master switch, the mode selector and the manual start pushbutton. The engine master switches and the mode selector are located behind the thrust levers. The manual start pushbuttons are located on the overhead panel.

The system can be used in the automatic start mode and the manual start mode. During the routine flight operation the automatic start mode is the normally used mode. The manual mode is used if the automatic mode must be overridden to prevent the automatic interruption of the start sequence.

7.3.3.1 Ground Start, Automatic Mode

At first the mode selector is turned to the position START for the selection of the start mode. This causes the EIUs to switch on the power supply for the EECs and the exciters. The electrical power for the ignition exciters is present at the ignition relays within the relay box that is installed below the EEC on the engine fan case. The ignition relays are controlled by the EEC. The next action is to move the engine master switch to the ON position. This releases the following steps of the start sequence:

- Opening of start valve
- Activation of one ignition exciter by closing its ignition relay between 10 and 16 % N2
- Release of the fuel flow at 22%N2
- Closing of the start valve and ignition off at 43%N2
- Acceleration to idle at 60%N2

The EEC uses the exciter that is assigned to the channel in command. Thus at the next engine start the other exciter is used. Over the complete start sequence the pilot can terminate the engine start by moving the engine master switch to OFF. After the last engine is started, the mode selector is turned manually to NORM.

7.3.3.2 Ground Start, Manual Mode

The manual start sequence also begins with the turn of the mode selector to START. But the next step is to press the manual start pushbutton. Due to this input the EEC opens the start valve only and waits for the next input. After the engine has reached an N2 speed high enough for light-off, the pilot moves the engine master switch to the ON position. The reaction of the EEC is to release the fuel flow and to activate both ignition exciters. The closure of the start valve at 43% N2 and the acceleration to idle is controlled automatically by the EEC.

7.3.3.3 Continuous Ignition

For the control of manual continuous ignition the mode selector is used. The selector position used for start selection is the ignition position for a running engine. Therefore it is labeled IGN/START. After the mode selector has been turned to the ignition position, the continuous ignition is activated on the running engines. Both exciters are powered for continuous ignition.

The continuous ignition can also be activated automatically by the EEC. This will happen under the following circumstances:

- Engine running and air intake cowl anti–icing is selected to ON
- EIU failed.
- Take–off or during flexible take-off
- Approach idle selected.
- In flight, when there is an engine flameout or stall detected by the EEC
- Reverse power selected

7.3.3.4 Limit Protection by the EEC

The most important incidents the EEC detects in the automatic start mode are a hot start, a wet start, a weak N2 acceleration, a faulty start valve and a compressor stall. If an EGT limit exceedance is imminent, a compressor stall or a wet start occurs, the EEC terminates the start sequence. This protection function with start sequence termination is active below the starter cutout speed.

If a limit exceedance is imminent in the manual start mode, the EEC detects this, too. It sends a warning to the ECAM display. A termination of the engine start by the EEC is not possible. This must be done by the pilot by moving the engine master switch to OFF.

For more information about unsuccessful starts see Appendix C.

8 Thrust Reverser Systems

8.1 Operation of Thrust Reversers

8.1.1 Basic Principle

To ensure a good braking effect for the aircraft also on contaminated runways (with water or slush) and for the reduction of brake wear, transport aircraft are equipped with thrust reversers. A thrust reverser allows the generation of a rearward-directed thrust force when deployed. To achieve this it redirects the exhaust gas flow at an angle of approximately 120 degrees. Figure 8.1 shows the direction of the airflow during reverse thrust operation. On turbofan engines with high bypass ratios, only the secondary gas flow is redirected by the thrust reverser because this gas flow generates the larger portion of the engine thrust. This results in a reverse thrust force high enough for braking purposes. To redirect the secondary gas flow only makes mechanical deflector components in the hot gas flow unnecessary. This results in a simpler reverser kinematics with less weight and costs.

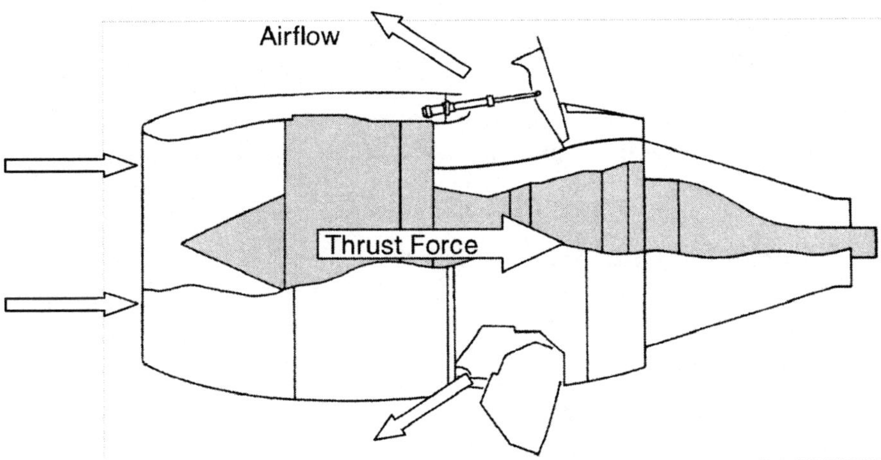

Fig. 8.1 Direction of the secondary airflow for the generation of the reverse thrust force

Because the effect of the reverse thrust is independent from the tire friction, the thrust reverser ensures a good deceleration of the aircraft on a contaminated runway with reduced tire friction.

During normal runway conditions the use of the thrust reverser requires less use of the wheel brakes for the same aircraft deceleration. The results are less wear of the wheel brakes and a longer operating life of the brake disks.

To have these advantages the higher weight of the engine nacelle must be accepted. To minimize the weight penalty caused by a thrust reverser the designers use a high percentage of composite material for the reverser and nacelle structure.

Usually all engines of an aircraft are equipped with a thrust reverser. But this is not a general rule in any case. An exception is the A380. For weight reduction purposes this aircraft has only two thrust reversers. They are installed on the two inboard engines.

8.1.2 Reverser Operation

A thrust reverser system is designed for the use on ground only. The system is equipped with safety features preventing the deploying of the reverser during flight. During landing the thrust reverser is deployed shortly after touchdown by selection of the pilot. The best braking effect is achieved at the higher speeds during the landing run because the propulsive efficiency for the reverse thrust has its highest values at the high forward speeds. With the decreasing aircraft speed it also decreases.

During typical flight operation the thrust reverser is used down to a speed of 80 knots. The pilot selects as much thrust as needed. This procedure ensures an operation with the highest propulsive efficiency and prevents the ingestion of dirt at slow taxi speeds. It is the most efficient way of thrust reverser use in terms of fuel consumption and brake wear.

8.2 Types of Thrust Reversers

Thrust reversers can be differentiated by the types of their subsystems. These subsystems are the

- Airflow deflection system
- Actuation system
- Control system

The airflow deflection system comprises the structural components necessary for the deflection of the airflow during the operation in the reverse thrust position. For the change between the forward thrust operation and the reverse thrust operation some components of the airflow deflection system are movable. To achieve its movement an actuation system is installed. This is controlled by the pilots via the thrust lever and the reverser control system. It is designed to move the reverser components into one of the two end positions. These are the forward thrust (stowed) position or the reverse thrust (deployed) position.

8.2.1 Airflow Deflection Systems

For the airflow deflection systems different designs are used. On large turbofan engines mainly two designs are utilized. These are:

- The cascade type reverser
- The pivoting door type reverser

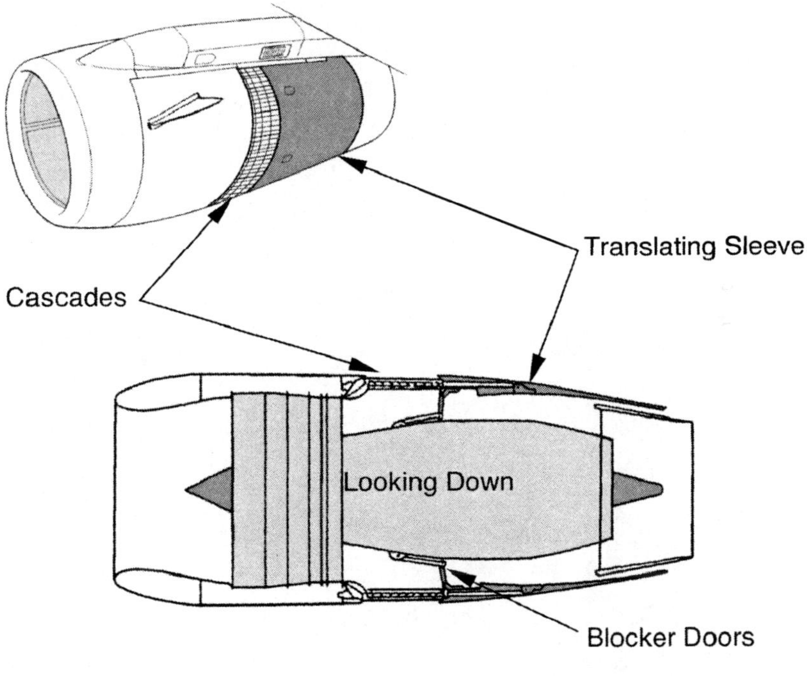

Fig. 8.2 The cascade type reverser of the V2500-A5 on an A320 in the reverse thrust position. The translating sleeves have moved to the rear and the blocker doors have pivoted into the secondary gas flow

The cascade type reverser has movable translating sleeves, which slides rearward to uncover the cascades installed on the fixed structure around the secondary airflow. In Fig. 8.2 this principle of airflow deflection is shown. Blocker doors are installed on the translating sleeves. They are pivoting into the airflow while the translating sleeves are moving aft. In this position the blocker doors are blocking the airflow passage to the rear. Thus the secondary airflow of the engine exits through the cascades, which deflect and accelerate the airflow into its final direction. This type of reverser creates low actuator loads and allows a very precise control of the airflow during reverse thrust operation.

The pivoting door type reverser has four pivoting doors installed in the openings of the fixed structure. Figure 8.3 shows such a reverser of the Airbus A320. In the open position the pivoting door blocks the airflow passage to the rear with its rear portion. The whole door acts as a deflector for the airflow. This type of reverser has fewer moving parts and simpler kinematics compared to the cascade type. Additionally the required stiffness of the fixed structure can be achieved at a lower weight. It creates a higher drag in the reverse position and the actuator loads are higher compared to the ones of a cascade type reverser of the same size.

The pivoting door type reverser was first used on the Airbus A320. It is also used on the A340-200/-300 with CFM56-5C engines and on the A330 with Trent 700 engines. The cascade type reverser is used on more aircraft designs and therefore it is still the most common design.

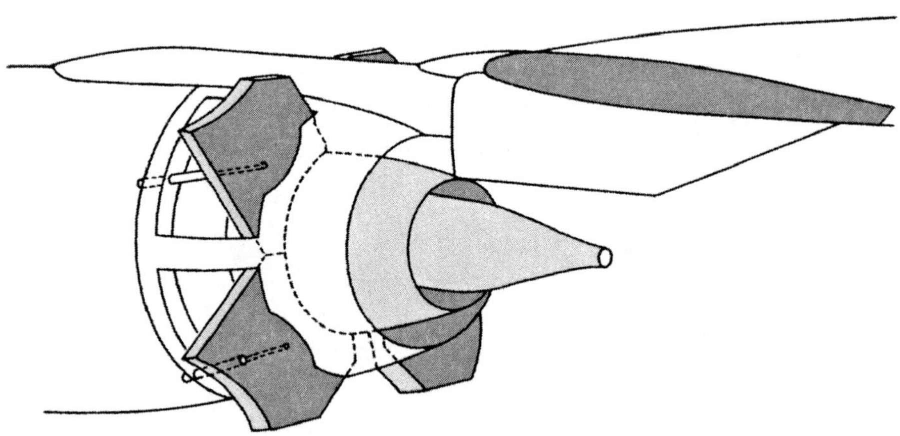

Fig. 8.3 The reverser of a CFM56-5A/-5B on an Airbus A320 with the four pivoting doors in the reverse thrust position

8.2.2 Actuation Systems

For the thrust reverser actuation of large turbofan engines two types of actuation systems are in widespread use. These are:

- The hydraulic actuation system
- The pneumatic actuation system

A hydraulic actuation system consists of several actuators, a control valve module and the locking devices. The actuators move the translating sleeves or the pivoting doors and are installed in the reverser structure. The hydraulic valves of the hydraulic control valve module are controlled by the reverser control system.

Figure 8.4 shows the actuation system components of the V2500 reverser. The control valve module of this system is installed on the underside of the engine pylon in front of the reverser structure. The control valves are controlled by the FADEC computer.

In the stowed position during the operation for forward thrust the reverser actuation system is depressurized. The movable parts are held in this position by mechanical locks. These locks are also components of the actuation system. They can be separately installed items, integrated into the hydraulic actuators or both designs are used in the same system.

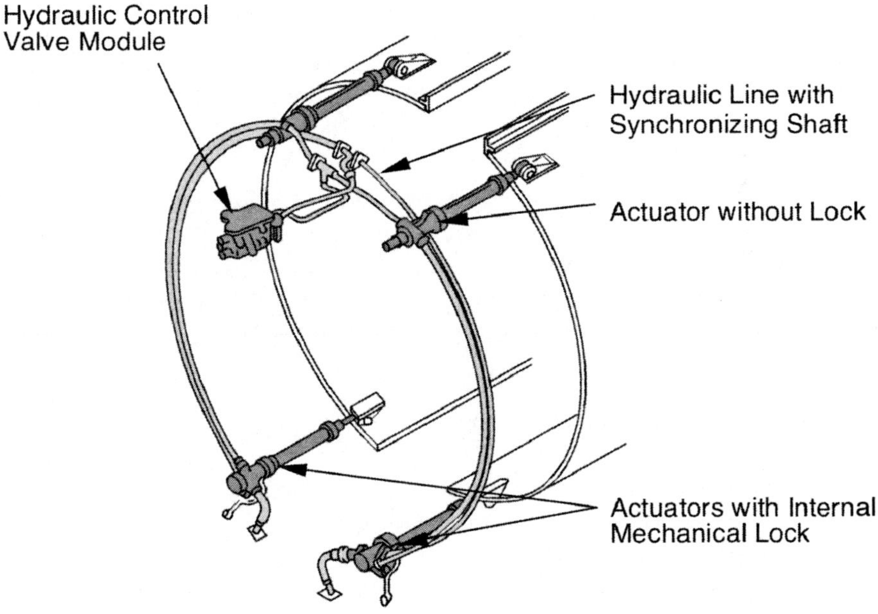

Fig. 8.4 The hydraulic actuation system of the reverser for the V2500-A5

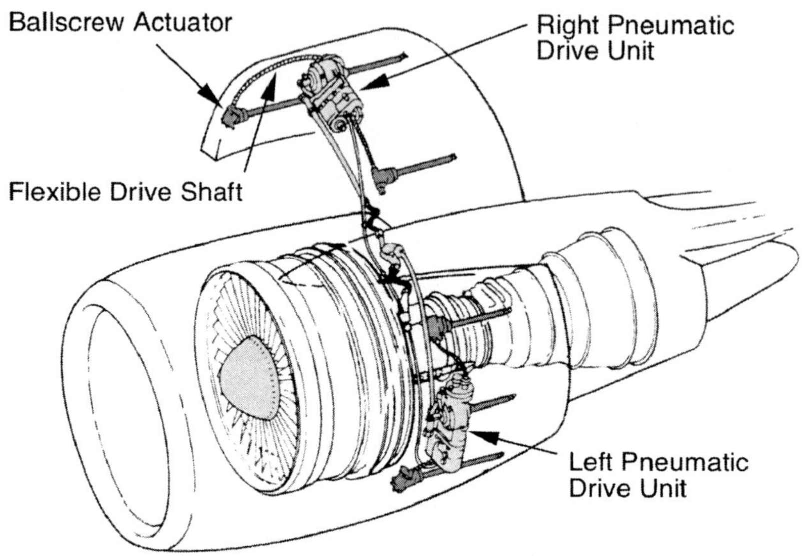

Fig. 8.5 The components of the pneumatically operated reverser for the CF6-80C2 used on the A300-600

The pneumatic actuation system consists of one or two air motors driving ball screw actuators via flexible shafts. The air motors are supplied with air from the high pressure compressor of the engine. The ball screw actuators move the translating sleeves. This type of actuation system is used on cascade type thrust reversers only. For the control of this actuation system the control system operates the installed air valves for air supply and directional control.

The locking function in the stowed position is achieved by a brake installed on the air motor. In most designs the air motor is integrated into one housing together with a gear, the brake and the directional control valve. Thus these components form the pneumatic drive unit of a reverser system. Figure 8.5 shows such a system used on the A300-600.

8.2.3 Reverser Control Systems

The used reverser control systems are either based on airframe mounted control circuits or on the reverser control logic of the FADEC computer. A reverser control system controls the switching of the hydraulic or pneumatic control valves and provides the necessary indication to the flight deck.

8.3 Reverser Structure

8.3.1 Fixed Structure

The cowling of a turbofan engine covers the core engine and also forms the gas path for the secondary airflow from the fan frame (or intermediate case) to the nozzle with its outer shell or barrel. It is designed in two halves and hinged to the engine pylon. Thus the cowling can be easily opened for maintenance access to the core engine. Due to the cross section of the cowling halves (aft looking forward) they are also called C-ducts. The airflow deflection system of the thrust reverser is integrated into the outer barrel of the cowling. Thus this cowling is often called the *reverser cowling*. The structure of the reverser cowling consists of the fixed structure and the movable structure.

The fixed structure is formed by the inner barrel, the forward frame, the longitudinal beams and, dependent on the design, parts of the outer barrel. The inner barrel has the function of the core engine cowling. The forward frame stiffens the whole structure together with the longitudinal beams and transfers the operating loads to the fan frame of the engine. It is attached to the fan frame with a positive-locking fit. Figure 8.6 shows the fixed structure of the V2500-reverser with the removed translating sleeves.

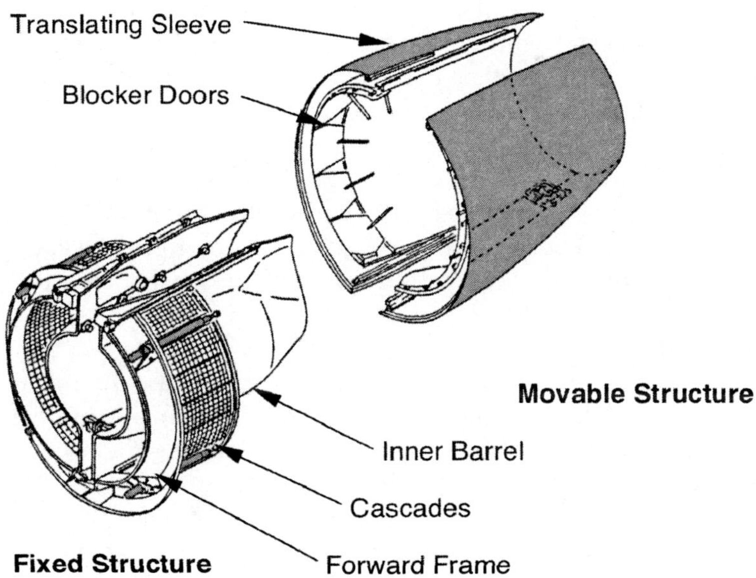

Fig. 8.6 The fixed structure of a cascade type reverser

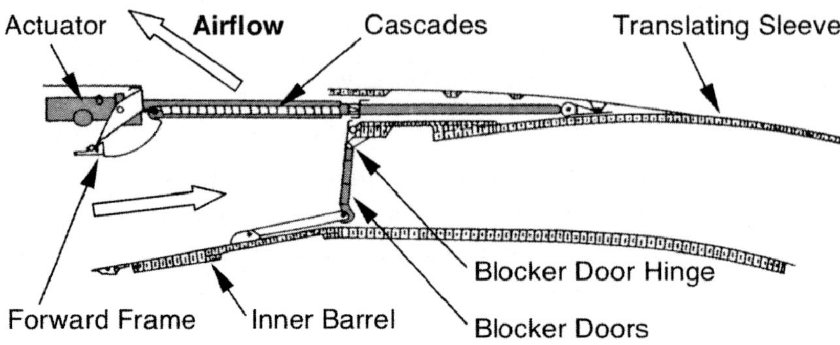

Fig. 8.7 The cross section of a cascade type reverser in the reverse position. The blocker doors have pivoted into the gas path of the secondary airflow. One of the actuators is visible. Its forward end is mounted to the forward frame

8.3.2 Movable Structure and Actuation System Components

The movable structure of a cascade type thrust reverser consists of the translating sleeves with the blocker doors. Figure 8.7 shows these components of a cascade type reverser in the reverse position. In the forward thrust position the translating sleeve covers the cascades and the blocker doors are positioned flush with the inner surface of the translating sleeve. In the reverse thrust position the cascades are exposed and the blocker doors have pivoted into the gas path to block the airflow to the rear. Thus the secondary airflow of the engine exits the gas path through the cascades into the direction enforced by the cascades.

On a pivoting door type reverser the four pivoting doors are the movable structure. Figure 8.8 shows the cross section of a pivoting door type reverser with a pivoting door in the reverse position. In the forward thrust position they cover the openings in the fixed structure. In the reverse thrust position they deflect the secondary airflow of the engine into the direction enforced by the door angle and the sharp edge (kicker plate) at the front end of the door.

The actuators are always mounted on the forward frame. If separate locks are installed, they are also installed on the forward frame. Usually the hydraulic lines and the flexible shafts connecting the actuators are installed on the forward side of the forward frame.

For the sensing of the blocker door or translating sleeve position switches are installed. On cascade type reversers LVDTs are used for the sensing of the translating sleeve position by the FADEC computer.

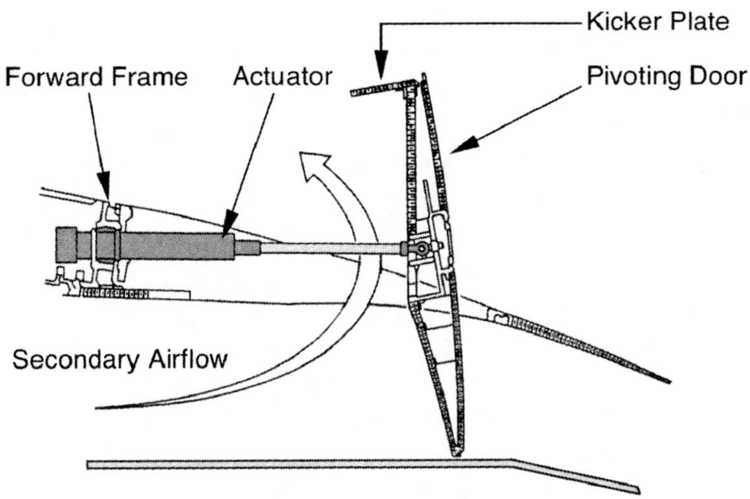

Fig. 8.8 Pivoting door type reverser of the A320. Each pivoting door has its own actuator

8.4 Reverser Control System

To introduce the reader to reverser control systems the systems of the B737 and the A320 are described as examples below. The comparison of the two will demonstrate the difference of the two philosophies very well. Both reverser systems are hydraulically operated.

8.4.1 Control System of the B737-600

The reverser control system of the B737 consists mainly of the reverser switches actuated by the reverse thrust lever and the relay circuits. The latter are representing the control logic of the system for the deploy and stow operation. Figure 8.9 shows a simplified schematic of this control system.

Each thrust reverser of the B737 is actuated by 6 hydraulic actuators. They are interconnected by flexible shafts for synchronization to prevent sleeve jamming. Without this synchronization the sleeves could jam by canting. The translating sleeves are locked in the stowed position by two locks in the upper actuators and two synchronizing locks. The latter prevent a rotation of the synchronizing shafts of the actuators and thus they also prevent a movement of the actuators. The synchronizing locks are electrically released. The hydraulic pressure on the actuators is controlled via the hydraulic control valve module. The air/ground signal, necessary to deploy the reverser, comes from the aircraft air/ground logic as an input to the reverser control logic. To deploy the thrust reverser the following conditions must be met:

- Deploy signal from thrust lever switches
- Hydraulic pressure and electrical power
- Aircraft on ground or less than 10 ft above ground

A deploy operation of the system is initiated by lifting the reverse thrust lever. This is possible only with the thrust lever in idle and triggers a deploy signal to the control logic. If the conditions for the deploy operation are met, the control logic switches the hydraulic valves into the deploy position. In parallel the synchronizing locks are released by the control system. These locks prevent an accidental deploy operation without a deploy command. This could happen if an electrical, hydraulical or mechanical system fault occurs.

The EEC senses the position of the translating sleeves. When they are fully deployed, the EEC releases the interlock solenoid of the reverse thrust lever. By lifting the reverse thrust lever further on the pilot commands more N1 speed with a decreasing thrust lever angle.

To stow the thrust reverser, the pilot lowers the reverse thrust lever fully to the stow position. The hydraulic valves are switched for the stow

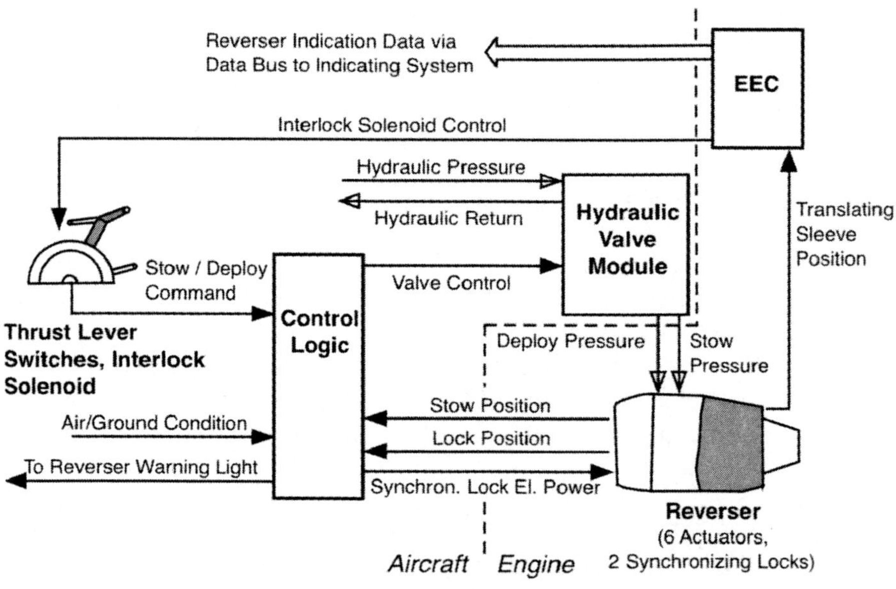

Fig. 8.9 Reverser control system of the B737-600 (simplified)

operation and the engine decelerates towards idle. When the translating sleeves have reached the stow position, the locks within the actuators engage. Shortly after the translating sleeves have reached the stow position, the control logic shuts off the hydraulic pressure and switches off the power for the synchronizing locks.

The data for the reverser position indication is provided by the EEC. It senses the translating sleeve position with the appropriate sensors. If a fault occurs during the reverser operation the control logic switches on the reverser fault light on the flight deck. This light also illuminates if an unlocked condition occurs during forward thrust operation (e.g. a primary latch in an actuator has failed).

8.4.2 Control System of the A320

The thrust reversers of the A320 family aircraft are controlled by the EEC of the respective engine. Figure 8.10 shows this in a simplified system schematic. This basic design principle is used on all Airbus aircraft with FADEC-controlled engines.

The thrust reverser of the A320 family aircraft with CFM56 engines is a pivoting door type thrust reverser. It has four door actuators and four mechanical locks that are released with hydraulic pressure. The EEC controls directly the hydraulic valves of the hydraulic valve module. It is also connected to the deploy and stow switches of the four pivoting doors. The hydraulic pressure for the reverser system is shut off upstream of the hydraulic valve module by an isolation valve. It opens only if the respective aircraft computer detects the thrust lever angle in the reverse range and the thrust lever of the other engine in the idle range. For the deployment the following conditions must be complied with:

- Thrust lever angle in the reverse range
- Thrust lever of the other engine near idle
- Engine running
- Hydraulic pressure available
- Ground signal from airframe systems

The reverser can only be operated with a running engine and the aircraft on the ground. After the thrust lever is placed in the reverse range, the EEC switches the hydraulic valves for the deploy operation. The hydraulic pressure releases the door latches first and then it enters each actuator. When the pivoting doors have reached the deploy position, the EEC closes the isolation valve within the hydraulic valve module to depressurize the system. The airflow keeps the pivoting doors open. During the deployment

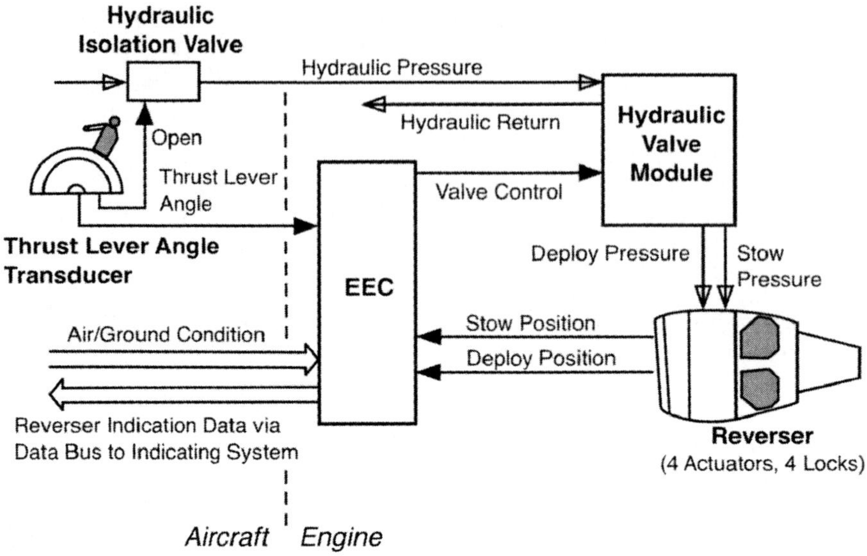

Fig. 8.10 Reverser control system of the A320 family aircraft (simplified)

of the pivoting doors the EEC keeps the engine at idle. After the EEC has detected all doors fully deployed, it accelerates the engine to the selected reverse thrust. During reverse thrust selection the thrust lever can be moved directly to the desired power setting, because it has no interlock stop. The interlock function for the delay of the engine acceleration is established in the software of the EEC. Thus no interlock stop for the thrust lever is necessary. This design keeps the thrust lever kinematics simple.

After the thrust lever has been moved into the forward thrust region, the EEC stows the pivoting doors completely and switches off the hydraulic pressure. If now all doors are locked correctly, they remain in the stowed position. Otherwise an unlocked door would be lifted a few degrees by the elastic seals and the air pressure. This would change the stow switch into the unlocked stage and the EEC pressurizes the system again for stowing the door. During the time the EEC senses an unlocked door, it sends an unlocked warning to the flight deck.

During the reverser operation the EEC sends the reverser position data to the indication system for the flight deck indication. If the EEC detects a system fault in the reverser system, it sends the corresponding fault message to the warning system.

The worst system fault would be a latch fault of the primary and secondary latch of a pivoting door during forward thrust operation. In this case the door would be opened by the airflow. The EEC reacts by commanding the engine to idle to minimize the drag of the open pivoting door.

9 Engine Anti-Ice System

9.1 Ice Build-Up on Engine Parts

When an aircraft flies through air with a high humidity and a temperature around the freezing point, moisture accumulates on the surfaces of the aircraft as ice. At the air inlet of a turbofan engine there are two areas where ice can accumulate on the surfaces. These are the inlet lip of the air inlet and the fan rotor. On the fan rotor the ice can accumulate on the fan blades and on the surface of the spinner. The ice on a surface of the inlet or the engine parts disturbs the airflow and generates vortices. In front of the compressor inlet, these vortices can lead to an unstable compressor operation. They facilitate stall and surge. Ice on the fan blades additionally produces imbalance, which leads to vibration. Detached ice pieces from the inlet structure create a risk for foreign object damage when they hit the fan rotor and the components behind the fan.

To prevent ice accumulation on the fan blades, no special installations are necessary. At N1 speeds above idle the centrifugal force sheds the ice from the fan blade due to the sleek surface of the fan blades.

The components located behind the fan rotor are normally in a warm environment due to the air temperature in these areas. Here ice will accumulate only at low shaft speeds. This can be a problem in the primary airflow for the operation of the compressors due to the vortices generated by the ice. As these components warm up during engine operation at higher shaft speeds, no steps are taken to prevent icing in these areas.

The methods used to prevent the accumulation of ice on the surfaces of the inlet lip and the spinner are described in the following sections.

9.2 Air Inlet Anti-Ice System

9.2.1 General

On the surface of an air inlet the ice starts to build up around the stagnation point of the inlet lip. To avoid this process it is sufficient to heat the inlet lip only. This can be done with warm air or electrically. On turbofan engines the usual principle is heating with warm air. To minimize the extraction of energy from the engine and to prevent the overheating of the inlet structure, the supply of warm air can be discontinued by closing the shut-off valve of the system. The pilot activates the system when he assumes that icing is imminent. On some aircraft ice warning systems are installed to give the pilots a hint about the icing situation.

9.2.2 System Components of a Pneumatic Anti-Ice System

For the supply of warm air a duct is installed from the HPC to the air inlet. The compressor stage for bleed air supply is selected by the designer in such a way that the warm air has the right temperature level for the anti-ice system during engine operation.

Upstream from the point where the duct is attached to the air inlet structure the shut-off-valve is installed. Usually pneumatically actuated valves are used. They operate with the air pressure from the anti-ice duct or with the air pressure from a more rearward HPC stage. The valves used can be simple open/closed valves or valves with a pressure regulator.

Within the air inlet structure the anti-ice duct is connected to a spray tube that is installed circumferentially behind the skin of the inlet lip. The air exits the spray tube through holes and impinges against the inner surface of the inlet lip skin. The warm air exits the structure through an outlet near the 6 o'clock position.

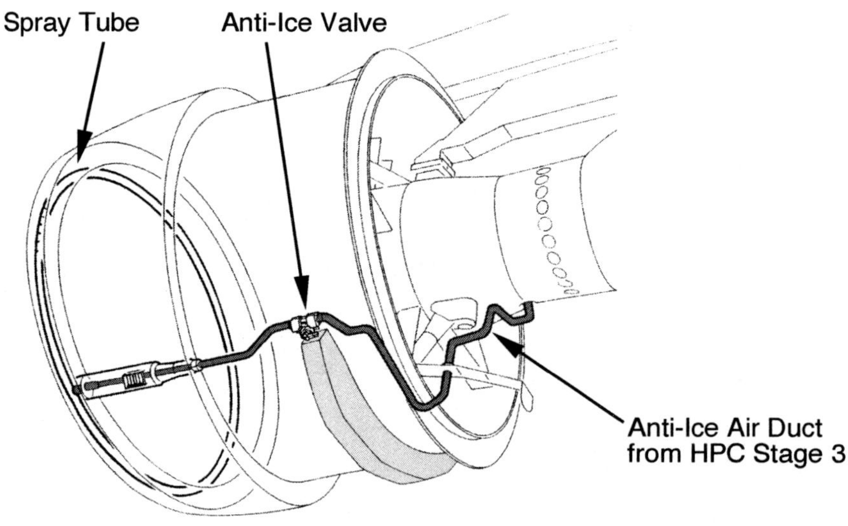

Fig. 9.1 Inlet anti-ice system component locations on an RR Trent 700

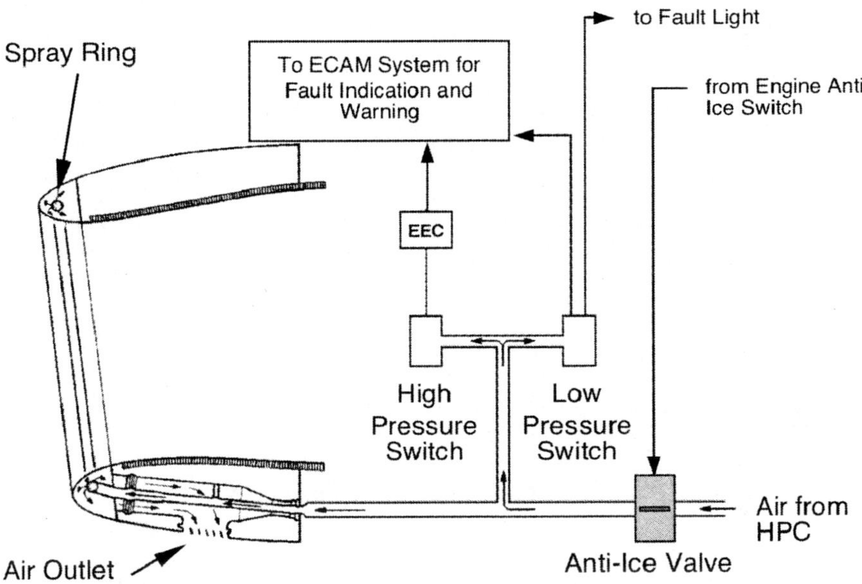

Fig. 9.2 Anti-ice system schematic of an RR Trent 700

The shut-off valve is controlled with a switch in the cockpit. Additionally a position indication and a fault indication are installed. These indications are usually accomplished by lights within the anti-ice pushbutton and by messages on the ECAM or EICAS display. Fault indication is used to announce a wrong position of the valve and in some systems it also warns of overheating and of overpressure downstream of the valve.

9.3 Spinner Anti-Ice

The tendency for the accumulation of ice on the surface of a spinner depends on the shape of the spinner. The typical spinner shapes used today are the elliptical, the conical and the coniptical spinner.

On an *elliptical spinner* the ice accumulation starts at a lower humidity level compared to the other spinner designs. For these spinners the heating with warm air is often used, because spinners of this shape have a higher risk for ice accretion. One example of this spinner heating is the PW4000. On this engine warm air from the LPC enters the N1 shaft and flows forward into the spinner cavity. The airflow exits the spinner through holes located just in front of the rear edge of the spinner.

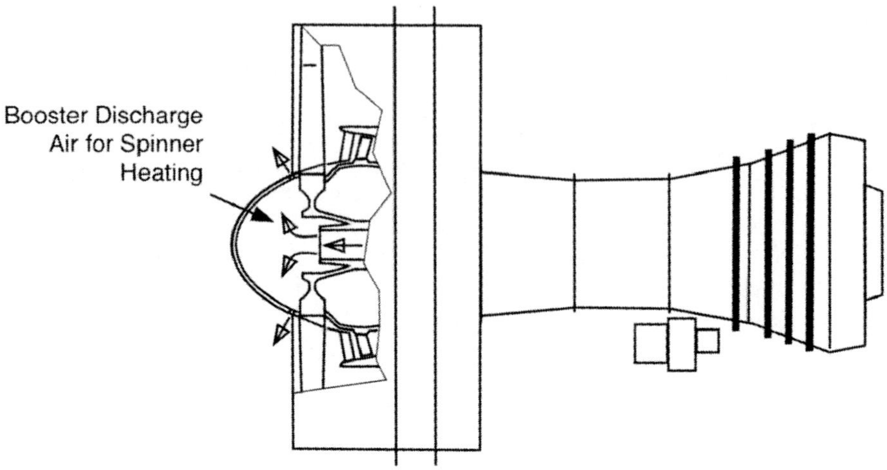

Fig. 9.3 Principle of spinner heating

A higher level of humidity is necessary for ice to begin to accumulate on a *conical spinner*. The reason for this is, that on a conical spinner the surface area at the stagnation point is nearly zero and the airflow moves almost parallel to the surface of the cone. On such a cone with a sharp tip icing starts at a higher humidity because icing always starts at the stagnation point of a component in the airflow.

A conical spinner is aerodynamically not as efficient as the elliptical one. But ice accumulation is a rare problem during operation because the conditions for ice accumulation on this spinner type are worse. Thus ice accumulation on the spinner can almost totally be prevented by the designer by selecting the conical spinner.

Fig. 9.4 Conical spinner on a CFM56-5A (© LTT)

Fig. 9.5 Coniptical spinner on a CFM56-5C (© LTT)

Both positive properties of the conical and the elliptical spinner are combined in the *coniptical spinner* as used on the engines of CFM and GE Aviation. The shape of this spinner is a combination of the conical and the elliptical spinner. It has a conical tip and a cambered rear cone. This shape combines the low tendency for icing of the conical spinner and the good aerodynamical efficiency of the elliptical spinner.

As stated above, ice build-up on a conical spinner is possible at high humidity levels with the right air temperatures. To prevent this ice build-up under these rare conditions, Rolls-Royce has developed a very simple design feature. On the conical spinners of the Rolls-Royce engines (V2500 and BR700 series as well) the spinner tip is made of an elastic rubber. The basic discovery for this design is the fact that the ice build-up on a rotating component is not symmetrical to the axis of rotation. When the ice build-up begins unsymmetrically on the elastic tip, the tip is deformed by the centrifugal force of the ice. Thus the ice flakes off and the ice build-up on such a spinner cannot spread behind the elastic tip.

Fig. 9.6 Conical spinner with rubber tip on a V2500-A5 (© LTT)

10 Power Plant

10.1 Nacelle

The complete propulsion unit installed on an aircraft is called a power plant. It consists basically of the bare engine and the components forming the nacelle together with the engine. These components are the air inlet, the fan cowls, the thrust reverser and the exhaust nozzle. They are brought together with the engine during engine installation to the aircraft. The fan cowl halves and the reverser halves are usually hinged to the engine pylon and remain attached to the pylon during an engine removal. The air inlet is attached to the front flange of the fan case. After engine removal it is swapped from the removed engine to the replacement engine.

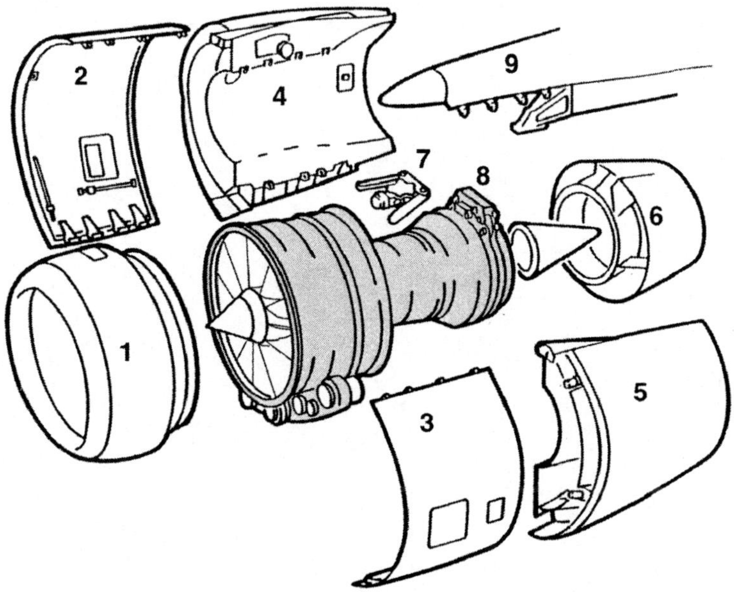

Fig. 10.1 Nacelle components and engine mounts of a V2500-A5. 1 Air Inlet, 2 Right Fan Cowl, 3 Left Fan Cowl, 4 Right Reverser Cowl, 5 Left Reverser Cowl, 6 Common Nozzle Assembly, 7 Fwd Engine Mount, 8 Aft Engine Mount, 9 Pylon

The exhaust nozzle is attached to the turbine frame of the engine and can be swapped between the engines or is delivered with the replacement engine. The fan cowls cover the fan case area of an engine and join the core engine cowlings with its secondary airflow nozzle. On most engines the thrust reverser is integrated in this structure. Thus this cowling is called the reverser cowling or reverser structure. It is typically designed in two halves. In Fig. 10.1 the nacelle components of a V2500-A5 are shown.

10.2 Bare Engine

The engine as delivered by the engine manufacturer is called the bare engine. Before it can be installed it needs some additional hardware to become a *demountable engine* (or *dressed engine*) for the installation to an aircraft. The process that changes the bare engine into a demountable engine is called *power plant build-up*. This build-up is usually done after the test cell run of the freshly assembled engine at the location where the engine shall be installed to the aircraft. Figure 10.2 shows a CFM56-5A as a bare engine and as a dressed engine. During installation to the aircraft the engine joins the other components necessary to become a power plant.

The following groups of components are added to an engine at different stages to get a power plant:

- During power plant build-up (PPBU)
  - Exhaust nozzle
  - Components to connect the respective airframe systems to the engine or to an accessory unit
  - Engine mounts
- During installation to the aircraft
  - Air inlet
  - Cowlings

10.3 Power Plant Build-Up

The components installed during power plant build-up on an engine are called the quick engine change components or *QEC components*. The term QEC has its origin in the fact that a prepared replacement engine with installed QEC components can be installed very quickly after the removal of the installed engine. No exchange of components between the two engines

Fig. 10.2 A CFM56-5A as a bare engine (left) and the fan case of a dressed CFM56-5A with the harnesses and the starter duct (right) (© LTT)

is necessary during the engine change. The QEC components are mostly not produced by the engine manufacturer. They are assorted to a kit that is delivered by the airframe manufacturer or a third partner company. If an engine type is usable for different aircraft types, different QEC kits are available to prepare engines of this type for the respective aircraft type. The QEC kit adapts the engine to the respective airframe system configuration. The actual make-up of the QEC kit will usually depend on the type of aircraft the engine will be used on. Typical engine types for a usage on different aircraft types are the CF6-80C, GEnX and the PW4000. The QEC kit may also be different for different engine positions on the same aircraft; in particular if fuselage mounted engines are used.

The QEC kit is made of a collection of components and accessories such as pumps, generators, wiring harnesses and fluid lines. In the engine pylon the fluid lines, the electrical harnesses and the pneumatic duct are segregated from each other by fireproof bulkheads. This segregation often continues on the engines. Thus all fluid lines are connected to the pylon on the same side and the components not carrying fluids like pneumatic ducts and electrical harnesses, are connected to the pylon on the other side.

The QEC kit mainly contains the components of the following airframe and engine systems:

- Engine Mounts
- Inlet Anti Ice System

- Electrical Power Generation
- Pneumatic System
- Hydraulic System
- Fuel Supply System
- Fire Detection and Extinguishing System
- Mechanical Engine Controls (if applicable)
- Engine Fluid Drains

10.3.1 Engine Mounts

For the attachment of the engine to the pylon the engine is equipped with a front and an aft engine mount. The engine mounts are the adaptors fitting to their counterparts on the engine pylon. They are installed on the bearing loads carrying engine cases. These are the compressor intermediate case and the turbine frame. The mounts transfer side loads, vertical loads and torque around the longitudinal engine axis. To make the path for the thrust force into the pylon as short as possible, the front mount additionally transfers the thrust force on most engines. Therefore the front mount is installed on the intermediate case directly above the HPC case. Figures 10.3 and 10.4 show this design used on a CFM56-5A.
On some engine designs the thrust force is transferred via the aft engine mount into the engine pylon. This requires the transfer of the thrust force from the intermediate case to the aft mount by two thrust links. The reason

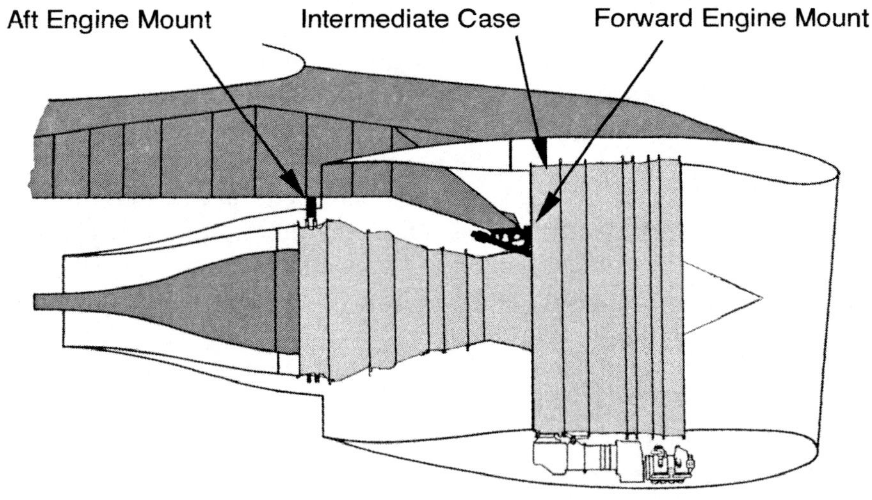

Fig. 10.3 The positions of forward and aft engine mount on a CFM56-5A installed to the pylon. The thrust is transferred via the forward mount

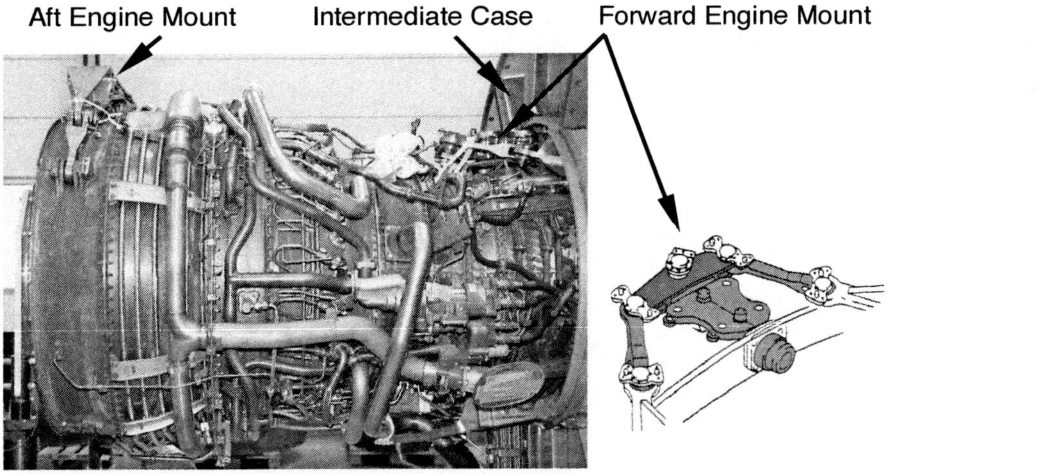

Fig. 10.4 The engine mounts installed on a CFM56-5A (© LTT)

for this design can be a lack of space on top of the engine for the forward mount or the design of the engine structure, as it is for the GE90. Due to the large diameter of the GE 90 fan case, a higher pylon structure would be necessary if the forward mount were attached to the fan hub frame above the HPC case. In Fig. 10.5 the arrangement of the engine mounts together with the thrust links is shown.

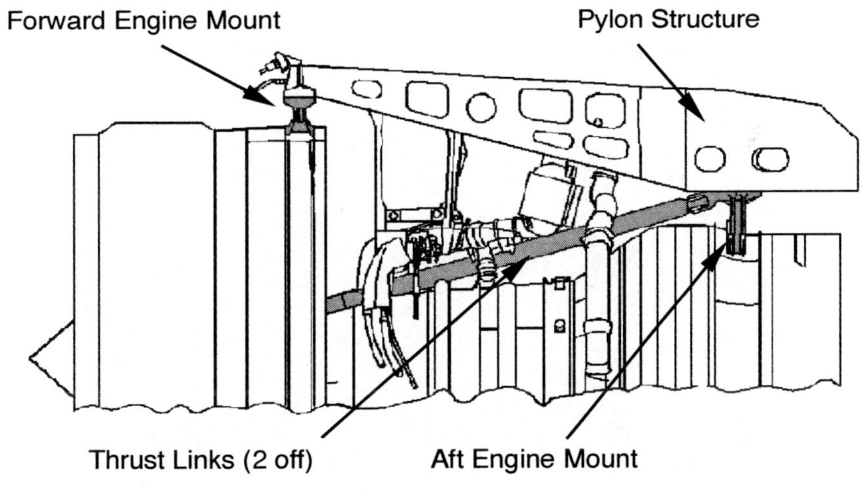

Fig. 10.5 Arrangement of the two engine mounts and the thrust links on a GE90 (© Boeing)

Because all forces transferred from the engine into the pylon are transferred via the engine mounts, these components are fail-safe designs. A complete failure of an engine mount or the thrust transferring components must be prevented because it would lead to the loss of the engine during operation.

10.3.2 Inlet Anti Ice System

See Chapter 9 for details.

10.3.3 Electrical Power Generation

For the generation of electrical power an integrated drive generator (IDG) or a variable frequency generator is installed with its wiring harness on the engine. The IDG or generator is installed on the assigned installation pad of the accessory gearbox. Some long range aircraft have two IDGs or generators on each engine. For the lubrication of the constant speed gear and the cooling of its generator the IDG has a closed circuit oil system with an external cooler. Air or fuel cooled coolers and installations with both types are used. The oil lines connecting the IDG with its oil cooler also belong to the IDG system and are installed during power plant build-up. For the electrical connections to the airframe the IDG harness with the power lines and the harness for the control system are installed on the engine case. The location of the IDG and its harness on a V2500-A5 are shown in Fig. 10.6.

Fig. 10.6 An IDG installed on the gearbox pad of a V2500 (© LTT)

A variable frequency generator also has a drive gear. But it has no constant speed function. In such a generator the oil is used for the lubrication of the gear and, like in the IDG, for the cooling of the generator. Therefore a variable frequency generator has an oil cooler as well.

10.3.4 Pneumatic System

The components of the aircraft pneumatic system, which are installed on the engine, remain installed during an engine removal. The pneumatic duct system on the engine is disconnected during the engine removal from the aircraft duct system above the engine where the pressure regulating valve is located. On the replacement engine all the components below this disconnect point are installed during power plant build-up.

The pneumatic system is connected to the high pressure compressor for the bleed air supply of the system at two compressor stages. One connection is called the high pressure bleed connection and the other is called the intermediate pressure bleed connection. The high pressure bleed connection is located on most engines at the rearmost HPC stage or at one of the rearmost stages. During cruise the pneumatic system is supplied through the intermediate pressure bleed connection. Thus it is located at an HPC stage that delivers sufficient pressure into the system. At power settings below cruise power the system is supplied through the high pressure bleed connection. This compressor stage can deliver a supply pressure that is high enough to keep the nominal system pressure even at idle.

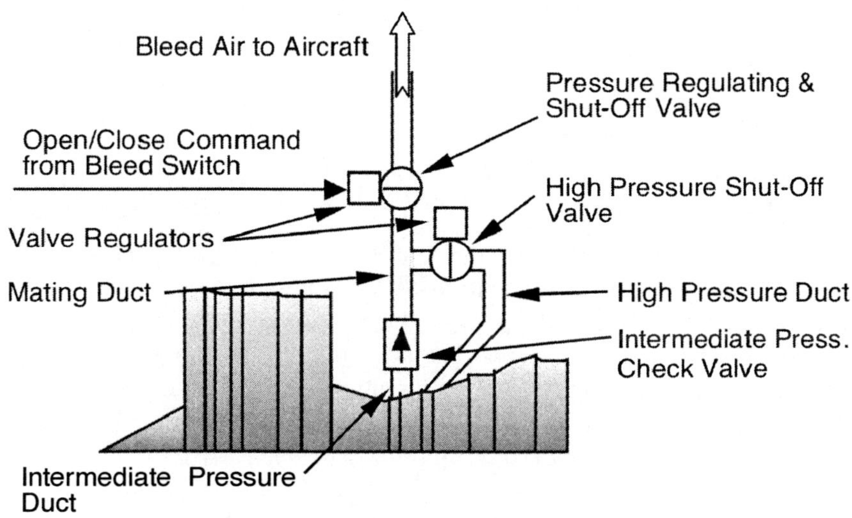

Fig. 10.7 The bleed air connections on the HPC of a CFM56-5A

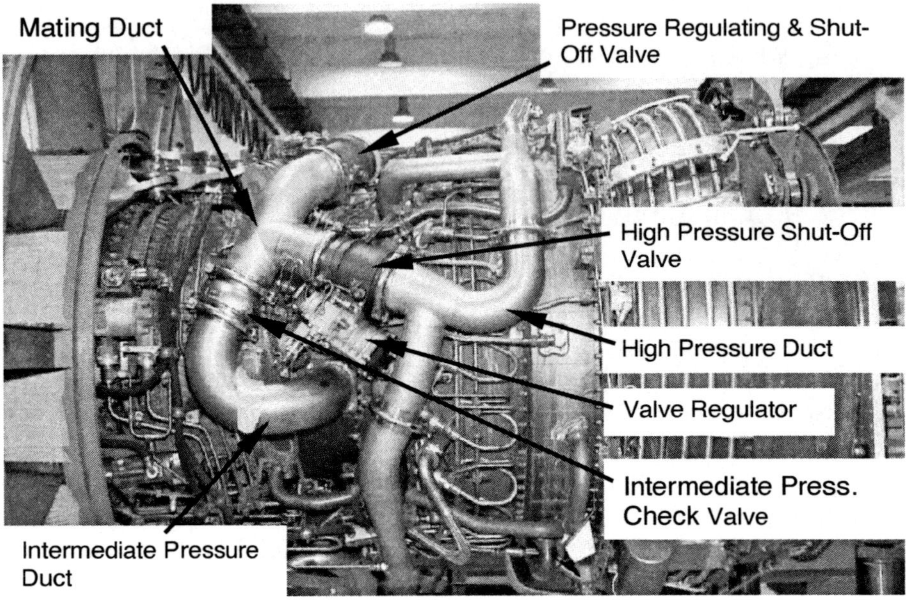

Fig. 10.8 The bleed air system on a CFM56-5A (© LTT)

On the CFM56-5A that is used as an example the pneumatic system is connected to the 5th HPC stage (intermediate pressure) and the 9th HPC stage (high pressure). Figure 10.7 shows the principle of the bleed air system connected to the high pressure compressor. In Fig. 10.8 the components of the bleed air system installed on a CFM56-5A are visible:

- High pressure duct (9th stage)
- Intermediate pressure duct (5th stage)
- High pressure shut-off valve
- Intermediate pressure check valve
- Mating duct segment
- Pressure regulating and shut-off valve

The two valves, the high pressure shut-off valve and the pressure regulating and shut-off valve, are equipped with pneumatic regulators. The high pressure shut-off valve controls the switching between the two air sources stage 5 and stage 9. The pressure regulating and shut-off valve controls the system pressure of the pneumatic system. It is also the shut-off valve for the bleed air supply into the pneumatic system and is operated via the engine bleed switch on the flight deck.

10.3.5 Hydraulic System

For the pressurization of the hydraulic systems of the aircraft one or two hydraulic pumps are installed on the accessory gearbox. Each hydraulic pump is connected to the airframe by three lines. Through the pump supply line the hydraulic pump is supplied with hydraulic fluid from the reservoir and through the pressure line the pump supplies the hydraulic system with pressurized fluid. Because the hydraulic pumps are axial piston pumps, hydraulic fluid leaks from the cylinders into the pump casing. This fluid is drained from the pump housing through the case drain line and carries the wear particles of the pump components. These particles are removed from the fluid in the case drain filter before it returns to the hydraulic reservoir. The case drain line is always the line with the smallest diameter of the three hydraulic lines. The upper ends of the hydraulic lines are equipped with quick disconnect couplings for the connection to the pylon. All these lines are visible in Fig. 10.9.

The hydraulic pumps, the three hydraulic lines for each pump and often the case drain filters are installed on the engine during power plant build-up.

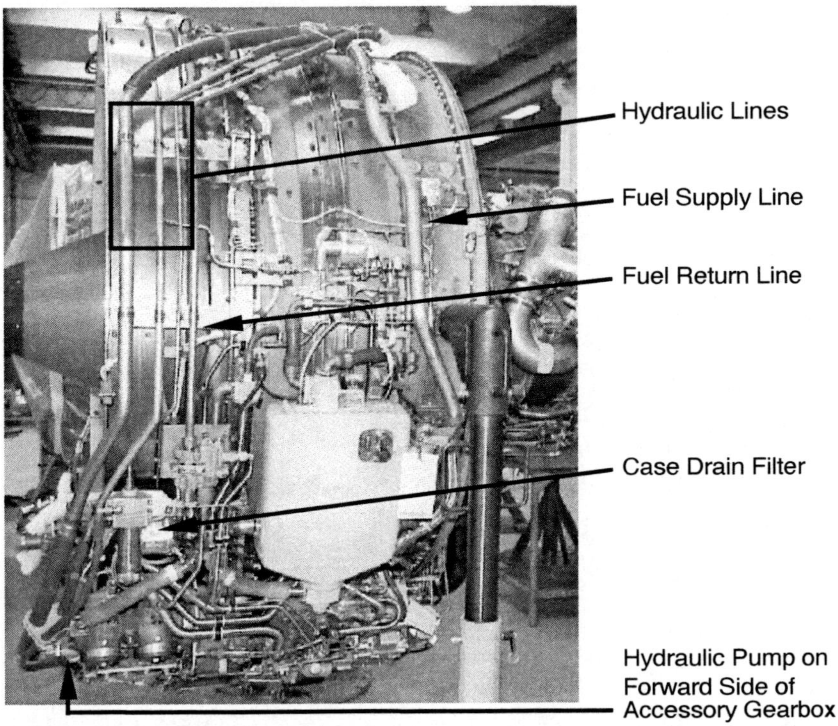

Fig. 10.9 Hydraulic system components and fuel lines on a CFM56-5A (© LTT)

10.3.6 Fuel Supply System

The fuel pump of the engine is installed on the accessory gearbox during engine assembly. The fuel supply line of the fuel pump is installed during power plant build-up. It connects the fuel pump to the fuel line connector on the pylon after engine installation.

On engines that have a fuel return function, the fuel return line is also added to the engine fuel system. Figure 10.9 shows both fuel lines installed on a CFM56-5A.

10.3.7 Fire Detection and Extinguishing System

The fluid lines installed on an engine contain flammable fluids. If such a fluid leaks out of a line or accessory component, there is a risk of fire in the well vented space between the cowling and the engine case. This would endanger the structural integrity of the nacelle and the pylon. To detect a fire as soon as possible, detectors for the fire warning system are located in the critical zones under the cowlings. If the accessory gearbox of an engine is installed below the fan case, the accessory zone under the fan cowling and the core engine/LPT zone under the core cowling are equipped with fire detectors. On engines with core engine mounted gearboxes all lines and accessory components containing flammable fluids are located under the core cowling. On these engines the fire warning detectors are installed under the core cowling only.

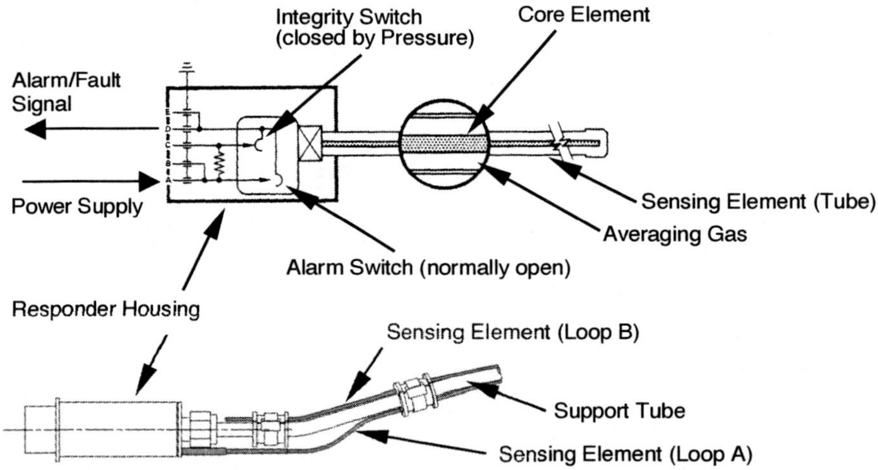

Fig. 10.10 An electro-pneumatic fire detector with its responder and the installation on the support tube (© LTT)

For the detection of an overheat condition created by leaking hot air from an engine case or duct, overheat detectors are installed. This can be the nacelle temperature sensor or the fire detector itself.

Detectors of two different types are used on turbine engines. One is the electro-pneumatic type, the other is the thermistor type. On most aircraft types the customer can select the fire warning system from the two types offered. The detectors are connected to their control unit, the fire detection unit, with an electrical harness. This control unit creates the fire warning for the flight deck when the air temperature around the fire detectors exceeds a set limit.

An electro-pneumatic detector element consists of a tube that contains an inert gas and gas emitting core, and has a responder on one end. The responder contains the pressure switches for sensing fire and integrity. In some systems the responder contains a second alarm switch for the detecting of an overheat condition. This switch closes at a lower temperature level than the fire warning switch.

The inert gas in a sensing element expands as a function of average gas temperature. The gas emitting core expels gas due to high localized temperatures. Both actions cause an increase in pressure in the element, which causes an alarm pressure switch in the responder to close, activating an alarm signal. Both actions are completely reversible as the temperature decreases. Then the pressure decreases, and the alarm switch deactivates.

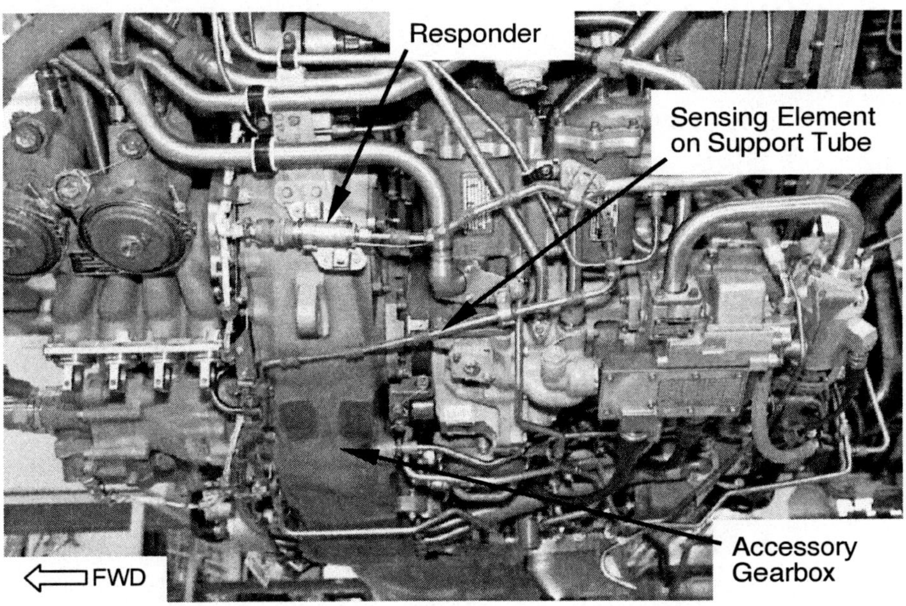

Fig. 10.11 Fire detector installation at the accessory units of a CFM56-5A (© LTT)

When a sensing element is damaged allowing the inert gas to leak, the integrity pressure switch opens, activating a fault signal. The pressure switches provide alarm and integrity signals to the appropriate detector control unit.

Figure 10.10 shows the principle of an electro-pneumatic detector element. In Fig. 10.11 the detector installation at the accessory units of a CFM56-5A is shown. The responder of one detector loop is visible. The fire detection system of this engine uses detectors at three locations. Additionally to the shown installation detectors are installed at the front flange of the LPT case and in the pylon structure above the forward engine mount.

A detector of the thermistor type is a temperature-dependent resistor (thermistor), whose resistance decreases as temperature increases. Each detector is a coaxial design, with the resistive material located between the inner and outer conductors. Or it is designed as a two-wire design with the two wires in the detector tube and the resistance between the two wires is sensed. An overheat warning or a fire warning is created if the resistance of the detector has dropped below a certain level.

The fire detection system is designed as a dual loop system. Thus two detector elements, one for each loop, are installed at each location.

Because the detector tubes of both detector types are very thin, they are installed with clamps on support tubes. Figure 10.12 shows the detector of a CFM56-7B.

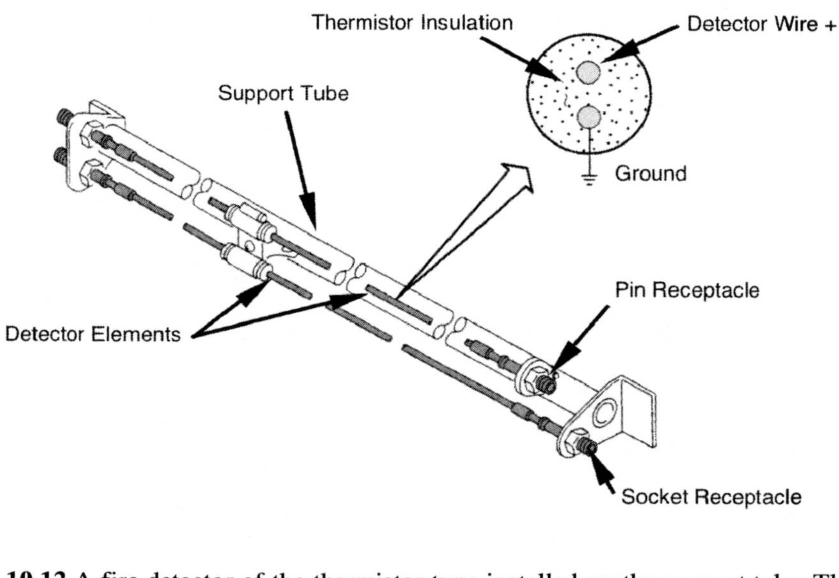

Fig. 10.12 A fire detector of the thermistor type installed on the support tube. The detector tube contains the two wires of the detector (© Boeing)

To extinguish an engine fire a gaseous extinguishing agent is used. Two bottles with this agent are installed in the engine pylon. After the release of the agent by the pilot it flows through tubes into the space under the cowlings and extinguishes the fire. The extinguishing tubes are often installed in the pylon only. On some engines extensions of these tubes are installed on the engine cases.

10.3.8 Mechanical Engine Controls

On engines with mechanical engine controls the control cables from the disconnect point at the engine pylon to the levers on the FCU are installed on the engine. These push-pull cables are mounted to the engine case with several clamps. The connection to the FCU levers is accomplished with fork ends or with rack and pinion drives.

10.3.9 Sensors and Harnesses

The electrical components located on the engine must be connected to the pylon receptacles. For this purpose the harnesses are installed on the engine during power plant build-up. On some engine types additional sensors for the engine condition monitoring are installed as well.

On FADEC engines most components are connected via FADEC harnesses to the FADEC computer or to the airframe circuits. Dependent on the engine type some additional harnesses may be necessary for the fire detectors and other functions not covered by the FADEC system.

10.3.10 Engine Fluid Drains

Where it is very likely that a flammable fluid can leak out of a component or line, the fluid must be drained away from the engine case and out of the engine cowling to prevent the build-up of a flammable fluid-air mixture in the cavity under the cowling. The flammable fluids present in the components on the engine are generally fuel, engine oil, hydraulic fluid and the oil of the generator system. The leaking fluids are drained through thin stainless steel tubes to an overboard drain mast located in the 6 o'clock position of the engine.

Typical areas where components may leak are the drive shaft seals of the gearbox-mounted units and the piston seals of fuel driven actuators. Thus drain lines run from the gearbox mounting pads and from the actuators on the engine to the *drain mast*.

It may also be possible that rainwater can accumulate in cavities of the nacelle and the engine pylon. To prevent this, these areas are also drained and some of these drain lines are also routed to the drain mast.

In very simple drain system designs the fluids from all drain lines can run through the drain mast overboard over the whole engine operating time- in-flight as well as on the ground.

More sophisticated systems provide *drain tanks* (or holding tanks) to catch the drained fluids while the aircraft is on the ground. The holding tanks are drained during the next flight by ram air pressure via the drain mast. Such a design prevents slippery puddles under parked aircraft.

Appendix A
Data Transfer in Digital Aircraft Systems

A.1 Serial Interfaces

A large number of sensors in the airframe and engine systems require only a low volume data transfer between the sensors, the microcontroller and the central processor. Because every computer depends on the communication with peripheral devices, it contains components for data transfer with external equipment. Prevalent on PCs as well as on industrial, scientific and consumer devices is the RS-232 (EIA-232) serial port as a means of control, monitoring and low volume data transfer. One port can connect to only one peripheral device.

A serial port transmits and receives data one bit at a time over one wire. While it takes longer to transfer data this way, only a few wires are required. Two way (full duplex) communication is available with only three separate wires, one to transmit, one to receive and a common ground wire. The RS-232 specifications include numerous additional control lines, which are used for special applications only.

Generally processors internally use 8-, 16-, 32-, or 64-bit parallel data buses for faster processing. Thus data intended for transmission on a serial data line has to be converted from a parallel to a serial data stream. Likewise data received has to be converted from a serial to a parallel data stream. These conversions can be performed by applying appropriate software, which is a method generally used on microcontrollers.

A specific piece of hardware that converts data between a parallel bus and an RS-232 interface in both directions is the UART (Universal Asynchronous Receiver-Transmitter). The RS-232 serial port is an asynchronous device. For an asynchronous transmission its start is identified by a start bit and the end by one and a half or two stop bits. The data bits are sent to the receiver after a start bit. Such a data character usually consists of 7 or 8 bits. A parity bit may optionally be transmitted after the data. The transmitter and the receiver must agree on the number of data bits and the transfer rate. After converting a character to be transmitted from parallel to serial, the UART adds the start and stop bits and sends the result to the serial port. Every character received is stored, the start and stop bits are removed and the character is converted from serial to parallel. Then it is ready to be read by the processor.

The UART usually does not directly generate or receive the external signaling levels (voltages) that are used between the devices. An interface is used to convert the logic level signals of the UART to the external signaling levels. All signals are measured in reference to a common ground. A positive voltage between 3 and 15 V represents a logical 0 and a negative voltage between -3 and -15 V a logical 1. This switching between positive and negative is called bipolar. The zero state is not defined and is considered to be a faulty condition (this happens when the device is not operating). The dead area between +3 and -3 V is designed to absorb noise. By employing the RS-232 specification, use is made of a mature and universally available PC technology.

If higher noise immunity, higher data transfer rates and more complex networks are necessary, one of two other serial interfaces is used. These are the (TIA/EIA) RS-422 and RS-485.

These use two twisted wires for the data transmission. On one line a true signal is transmitted and on the other an identical signal but of opposite polarity. This produces opposing currents and magnetic fields, thus minimizing the emitted electromagnetic interference by cross-canceling the opposite fields around each wire pair. Furthermore noise is coupled to both wires of the pair in the same way and thus is common to both signals. As the receiver evaluates the difference between the voltages of both wires, the effect of the noise is eliminated. This noise immunity allows for transmissions at higher data rates and over longer distances.

(TIA/EIA) RS-422 allows one transmitter to connect unidirectionally (simplex) to up to 10 receivers. Due to the lack of bidirectional capabilities allowing multipoint connections, the (TIA/EIA) RS-485 standard was created to add this capability. Its standards only describe the electrical properties of the network. No transmission protocol is specified in the standard.

However, several field bus standards in use specify RS-485 as electrical standard of data transmission. Among these are Profibus and Interbus-S. These widely used and proven standards can readily be adapted. RS-485 meets the requirements for a truly multi-point communication network. The standard specifies up to 32 transmitters and receivers. In a four-wire network one node is the master and all the others are slaves. The master node communicates to all slave nodes. All slave nodes communicate only with the master node. The slave nodes never listen to another slave's response to the master. The master commands a response from only one slave at a time and thus prevents data collision. Adopting the RS-485 interface allows to take advantage of the widely used industrial field bus technology, for which plenty of proven hardware and software is available.

A.2 Data Buses

A.2.1 ARINC 429

The ARINC 429 avionic data bus was specifically designed for use in civil transport aircraft and introduced in 1977. It defines how avionic devices and systems can communicate with each other. The specifications define the electrical and data characteristics and as well the protocols to be used. ARINC 429 is a unidirectional (Simplex) data bus using only one transmitter and at least one and not more than 20 receivers.

The data bus consists of a screened twisted pair of wires with the screens usually connected to ground at both ends. Like the RS-422 and RS-485 interfaces always identical signals but of opposite polarity are transmitted along the two wires. This secures the noise immunity already described. The modulation is also bi-polar but different to the RS-422 and RS-485 interface standards. A logic state 1 is represented by a high state, which after one half of the bit length returns to zero. A logic state 0 is represented by a low state, which after one half of the bit length returns to zero. This principle eliminates the need for a synchronization signal and thus the need for start and stop bits.

The transmission voltage is 10±1 V between the wires. If one wire is +5 V, the other is -5 V and vice versa. Data is transmitted serially in 32-bit words. Thus the transmitter is transmitting 32-bit words or the null state. Each 32-bit word is separated by an interword gap of four bit-times.

The ARINC 429 specifications define a high data rate of 100 kbps ±1% and a low data rate between 12.0 and 14.5 kbps. Each bus always operates either at high or low speed.

The transmitter converts parallel binary words into 32-bit words. The receiver will convert these 32-bit serial words into parallel binary words readable by the processor to which it is connected. Bits 1 through 8 of a 32-bit word contain a system address label (SAL). This label not only defines a parameter but also the data type and therefore the rule how the 23 of the remaining 24 bits have to be interpreted. Three protocols are defined in ARINC 429 standards. These protocols are for the transmission of numeric data, discrete data and data files.

The standard defines many fixed labels, which also define how the data is presented. Thus all manufacturers of different pieces of equipment always have a common basis on which data is presented. The receivers are programmed to look for only that data which is relevant to its operation. Relevant data is identified by the label.

The last bit is always the parity bit. Odd parity is specified. That means, that there must be an odd number of "1" bits in the 32-bit word. The parity bit is set according to the number of "1" bits.

Since the ARINC 429 data bus transmits data in only one direction it is extremely reliable. There is a very low probability of data conflicts or data corruption. Thus it has become the mainstay of federated avionics architectures.

A.2.2 ARINC 629

The ARINC 629 data bus was originally developed and patented by Boeing and integrated into the Boeing 777 design. In 1989 it was released as an ARINC specification. Cost concerns about this avionic specific bus system and the emerging popularity of Ethernet technology have prevented further ARINC 629 implementation after the B777 development. Nevertheless this standard initiated the move from the federated avionics architecture – supported by the ARINC 429 bus – to the higher levels of integration of the integrated modular avionics architecture.

ARINC 629 is a bidirectional distributed control bus capable of supporting up to 120 users. A 2 Mbps serial data transmission rate is specified for a twisted pair of conductors. Thus it is 20 times faster than an ARINC 429 bus. Each coupling to an ARINC 629 data bus is made via an ARINC 629 transmitter/receiver terminal. Data is transmitted along the bus serially in a self-synchronizing Manchester format. The terminal encodes transitions of "1" and "0" binary states rather than their static values. The transition induces an electromagnetic field around the wire. If the wire pair is fed through a bus coupler, which acts like a transformer, it will receive or transmit a signal. A connector is not required. Data communication between the terminal and the user is accomplished via a serial/parallel interface to a read/write memory shared between the terminal and the user system.

Information is transferred in 16-bit data words. To each word a parity bit and a three bit-times synchronization pulse is added. Each receiver listens to all data transmissions and selects from these the one it needs. ARINC 629 specifies the format of many data types including discrete, binary, real and graphic ones.

The physical network for bidirectional data exchange is simple, but the control is rather complex. The protocol must handle not only standardized messages, but also arbitrate data bus transmissions to ensure that only one terminal transmits at a time and that receivers listen at the proper time.

ARINC 629 defines a digital communication system where terminals connected to the bus may transmit and receive digital data using a protocol, which can be described as a carrier sense multiple access/collision avoidance. The bus access control is distributed along all the participating terminals. Only one terminal may transmit on the bus at a time. To avoid data collision, terminals listen to the bus traffic and start transmitting only when the bus is not busy.

Two protocols, basic protocol and combined protocol, are defined allowing transmissions to be either periodic, aperiodic or a combination of both. Basic protocol provides an environment where each terminal has equal access to transmit on the bus.

Upon starting to transmit its data, each terminal activates interval timers. These are programmed to prevent the terminal from transmitting data again before all other terminals have had the opportunity to transmit data. Three timers, called transmit interval (T1), the synchronization gap (S6) and the terminal gap (T6) are activated to secure efficient bus utilization and prevent data collision. The T1 value is determined by the minimum update rate requirement of the devices served. A terminal starts its T1 timer at the same time it starts transmitting and will not start transmissions again until T1 has elapsed. The synchronization gap (S6) ensures that all terminals are given access to the bus, while the terminal gap (T6) controls the sequence in which the terminals use the bus. When bus utilization reaches 100 %, the traffic will automatically slip into an aperiodic mode.

Combined protocol supports both periodic and aperiodic data transmission through the use of three data transmission priority levels. Periodic-level-1 transmissions are performed consecutively until all level-1 transmissions are completed. The time remaining between periodic transmissions is available for aperiodic level-2 and -3 data transmissions. A priority scheme ensures that any level-2 data is transmitted before level-3 data.

The B777 employs ARINC 629 for avionics functions and the safety critical fly-by-wire flight controls. A set of three ARINC 629 buses has been certified as a separate subsystem of the flight control system. These are also physically separated from the buses of the avionics functions.

A.2.3 AFDX

In the late 1990s investigations started to define a data network for the next generation of commercial aircraft based on the IEEE 802.3 Ethernet.

The intention was to take advantage of commercial-off-the-shelf (COTS) hardware and technology to reduce cost and development time.

Airbus Industries began work to define a new data network for use on the A380. Its Avionics Full Duplex Switched Ethernet (AFDX) is a special application of a network which is compliant with ARINC 664. Boeing uses the AFDX technology on the B787.

Avionics Full Duplex Switched Ethernet is a standard that defines the electrical and protocol specifications for the exchange of data between avionic subsystems. It provides much more functionality and is one thousand times faster than the ARINC 429 bus.

The avionic system of an airplane using the AFDX network in its integrated modular avionics architecture comprises the following components:

- Avionic subsystem
  An avionic module served by a computer. The computer contains an embedded end system that connects the avionics sub system to the AFDX network.
- AFDX end system
  The end system provides the interface between the avionic subsystem and the AFDX network. It furnishes a reliable and secure two-way data exchange with other avionic subsystems.
- AFDX network
  A full duplex switched Ethernet interconnect, consisting of a network that forwards Ethernet frames to their intended destination.

Full duplex switched Ethernet eliminates the possibility of data collision. Each avionic subsystem is directly connected to a switch via a link that consists of twisted wire pairs. One pair for transmission (Tx) and one pair for reception (Rx). The switch contains buffers and thus it is able to buffer multiple data packets in a first-in/first-out (FIFO) order for reception and transmission. The processing unit's function is to move data packets from the receive buffers to the respective transmit buffers. Thus it reads and interprets each arriving data packet next in line in the receiver buffers and determines its destination address.

After interrogating the forwarding table it determines, which Tx buffer or buffers are to receive the packet. This is then copied consequently into the Tx buffer(s). In first-in/first-out (FIFO) order the data packets are transmitted to the selected avionic subsystem or to another switch.

The network is wired in a star topology. Each switch connects to another switch. This switching architecture avoids collision but a data packet may experience a delay because it may have to wait until other data parcels have been transmitted.

An avionic computer connects to the AFDX network through an end system. The avionic computer is part of the integrated modular avionics architecture and is capable of supporting multiple avionic subsystems. Partitions isolate the subsystems using the same computer. These subsystems communicate with each other through communication ports. Thus end systems must provide communication interfaces for these ports. Each port has its own dual identifier.

An avionic subsystem may send a message through its communication port to its respective end system. Then the message is sent through the network to another end system where it is routed to the destination port of another avionic subsystem. One communication port can also send its message to several ports, which may be located in different end systems.

The units of data transferred across the network are defined as frames. The AFDX frame format contains the source and destination port address. The "payload" of the AFDX ranges from 17 to 1471 bytes. The frame size is variable and ranges between 64 and 1518 bytes.

The most important concept of the AFDX is the introduction of the virtual link. Each virtual link is a communication channel defined as a partition of the AFDX network. Similar to partitioning the access of an avionic subsystem to computer use, concepts are introduced to partition the access of virtual links to the same physical medium. This isolation of individual virtual links is secured by limiting the rate of transmission and the size of the frames. Thus a message may have to be partitioned and transmitted in several frames.

Each transmitting communication port is associated with a virtual link. Each message sent for communication receives additional information identifying source and destination end system addresses and source and destination communication ports. This frame is then placed into the appropriate virtual link queue for transmission. The transmission of the frame is scheduled by the end system's virtual link scheduler, which is also responsible for scheduling the transmissions of all virtual links originating from this end system. This scheduling to transmit information along several virtual links within a limited time frame may introduce "Jitter" . This may be explained by considering a multilane motorway reducing to a single lane. Jitter is a time delay whereby the rate of flow is reduced to attain a regulated arrangement of traffic. Jitter must be kept in set limits. By keeping frame rates, frame sizes and jitter under control a deterministic data transfer can be attained.

End systems communicate via two communication channels. All wiring and the switches are duplicated. This provides protection against the loss of one complete network.

Appendix B
Servo Valve Control for Actuator Positioning

B.1 General

In Chapter 5 the closed loop control principle for actuators on FADEC-controlled engines is described. For fuel pressure driven actuators the electrohydraulic servo valve is an important part of this closed control loop because they are used for the control of the fuel pressure to the actuators. The servo valves are located in the actuators or in the HMU of an engine.

The actuator force is generated from the fuel pressure supplied to the actuator. A design goal of the system is to control the fuel pressure to the actuators with as little electrical energy as possible. This helps to keep the demand for electrical power by the FADEC system low. Thus the servo valves are designed to be controlled with very low DC currents.

An electrohydraulic servo valve comprises the flow control valve and a torque motor. The electrical signal from the EEC for the control of the flow control valve (and actuator) is a DC current to this torque motor. A torque motor is a DC motor with limited armature deflection. This deflection is less than 1 degree and its actual value depends on the current from the EEC. The currents used are in the range of 100 mA.

Electro-hydraulic servo valves of different designs are used. In the following the three most used designs are described. The flow control valves of all these valves are moved by fuel pressure. The designs differ in the way the fuel pressure for the flow control valve movement is controlled:

- Hydraulically moved flow control valves with spill valve
- Hydraulically moved flow control valves with fuel jet nozzle
- Hydraulically moved flow control valves with pilot valve

B.2 Servo Valves with Spill Valves

Servo valves with spill valves consist of a spill valve for the control of the flow control valve, the flow control valve and the torque motor. The flow control valves control the fuel flow to the actuator. Two nozzles and a flapper form the spill valve. The flapper is located in the space between the two nozzle outlets. It is moved to the right or left of its neutral center position by the torque motor. Figure B.1 shows these details in a simplified schematic of this servo valve.

The two nozzles in the component and the flow control valve heads are supplied with the same servo fuel pressure. If the flapper is exactly in the center between the two nozzles, the fuel flow from both nozzles will be equal. In this condition the backpressure in both supply lines will also be equal and the flow control valve is in the central or null position. No fuel flows to or from the actuator. The servo fuel from both nozzles flows through a return flow line into the low pressure part of the fuel system.

The movement of the flapper towards a nozzle increases the backpressure in the flow line of this nozzle and the respective flow control valve chamber. In the flow line of the opposite nozzle the backpressure decreases and the fuel flow increases. These changes in fuel pressures cause the flow control valve to move towards the side with the lower pressure.

These changes in servo pressure are used to position the flow control valve to direct servo fuel pressure to one actuator side (head end or rod end). The direction of the flow control valve movement determines the

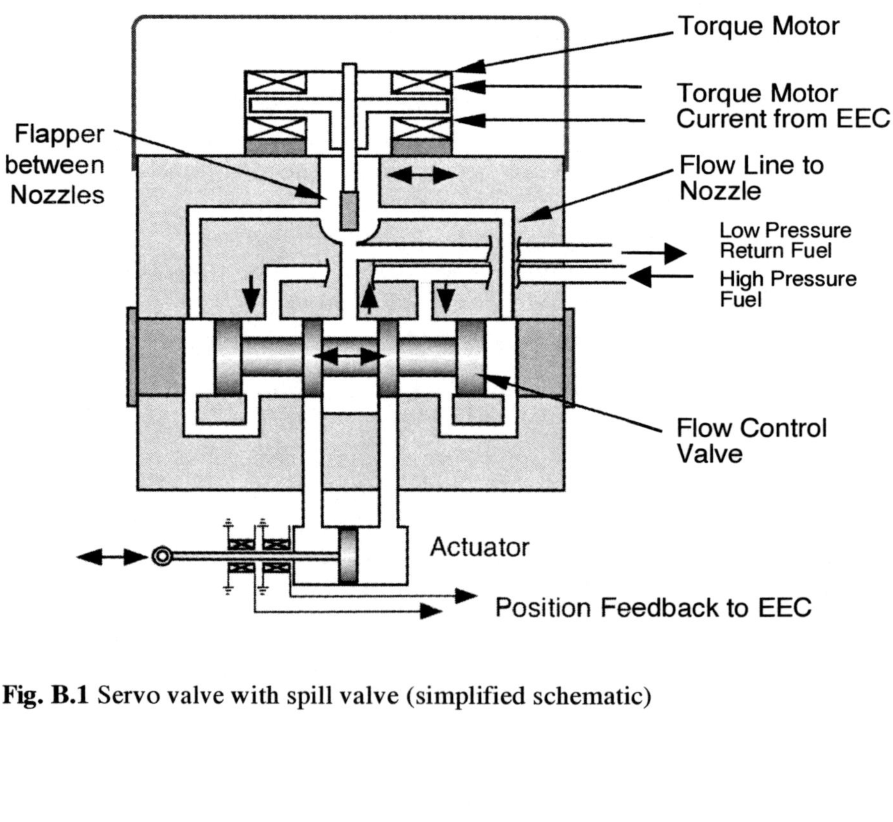

Fig. B.1 Servo valve with spill valve (simplified schematic)

moving direction of the actuator. The displacement of the flow control valve from the centered position determines the moving speed of the actuator.

This control of the flow control valve is possible with relatively small changes in pressure, because the control valves work smoothly and there are no counter pressures or forces to overcome.

When the actuator piston has reached the position commanded by the EEC (sensed with the electrical position sensor), a movement of the flapper in the opposite direction occurs. The EEC directs the flapper via the torque motor, in this case to the left.

The resulting change in servo pressures is varied as long as necessary to return the pilot valve to its null position. After reaching the null position the fuel flow to the actuator ends and the actuator remains in the current position.

B.3 Servo Valves with Fuel Jet Nozzle

In servo valves with a fuel jet nozzle a thin high pressure fuel jet is directed onto openings of the receiver flow channels, through which the fuel is guided to each side of the flow control valve. The nozzle position is controlled by the torque motor and is a function of the torque motor input current. Figure B.2 shows the design of a servo valve used on the CFM56-7B.

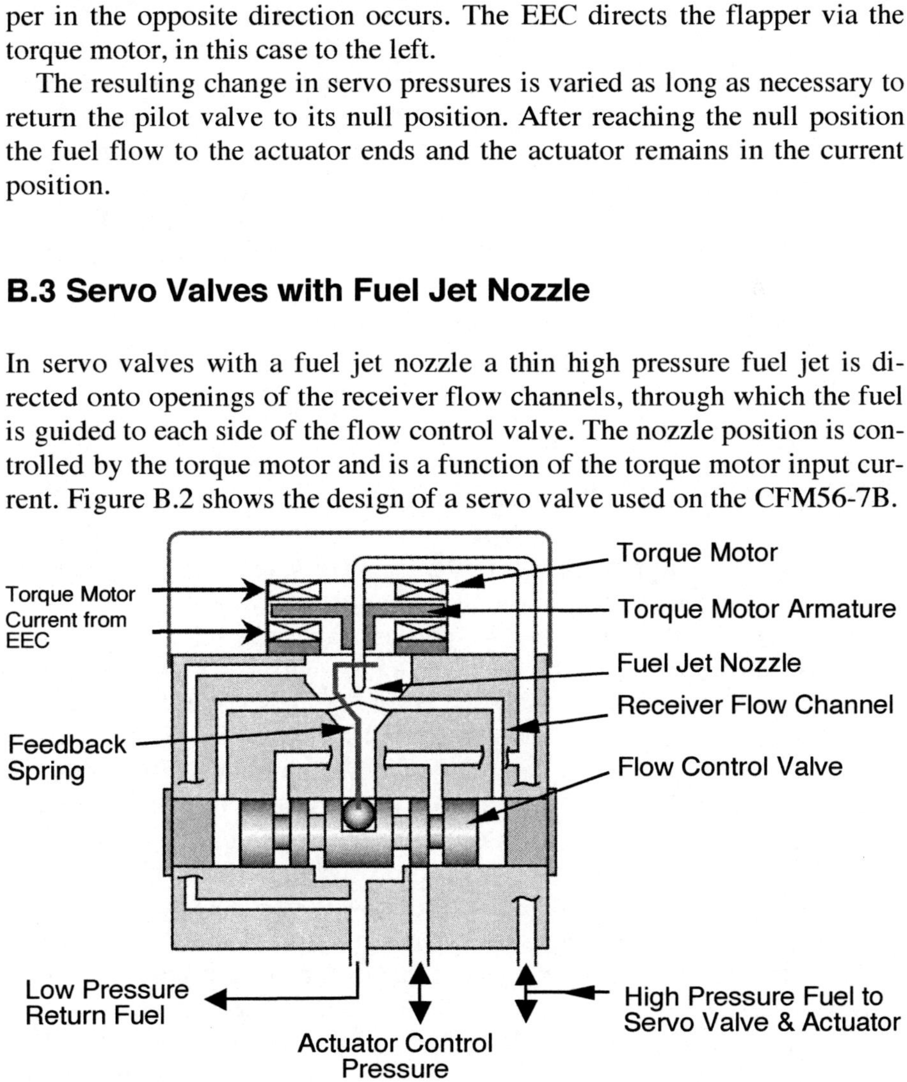

Fig. B.2 Servo valve installed in the HMU of a CFM56-7B (simplified)

When the torque motor positions the nozzle in the neutral position, equal pressure is supplied to each end of the flow control valve. An increase of high pressure fuel flow to one side of the flow control valve leads to its movement.

In the neutral position the valve land closes the case opening and no fuel flows to the actuator. Thus the actuator remains in its position. Each change in the flow control valve position changes the flow channel connections to the actuator. One land of the flow control valve directs the high pressure fuel to the actuator or the fuel from the actuator to the fuel return. The other side of the actuator piston is supplied continuously with HP fuel.

Every change in the torque motor current causes a deflection of the armature. The dimension and direction of the deflection is determined by the extent and polarity of the change from the null position current. The movement of the torque motor armature positions the jet nozzle.

The movement of the flow control valve is transferred back to the armature of the torque motor via a feedback spring. The lower end of the spring is connected to the center of the flow control valve. This mechanical feedback ensures that the armature, the nozzle and the flow control valve return to the null position faster after the torque motor input current is set to the null current value.

B.4 Servo Valves with Pilot Valves

In servo valves with pilot valves this pilot valve is used to control the fuel pressures for the movement of the flow control valve. The pilot valve is directly moved by the torque motor. The movable armature of the torque motor has an arm to move the pilot valve mechanically. Figure B.3 shows this arrangement, which is used in the HMUs of the CFM56-5 engines.

The pilot valve controls the valve control pressures for the flow control valve and causes the positioning of the flow control valve by means of these pressures. When the flow control valve is activated, the pilot valve does not move with it because it is connected to the torque motor armature. The axial movement of the two valves is always in the same direction. The flow control valve controls the servo fuel flow to and from the actuator.

When the flow control valve is centered, the flow to the actuator is zero. To keep the servo valve in this position the torque motor must balance the spring force acting on the pilot valve. The current necessary to center the servo valve is called the null current. For a movement of the actuator the servo valve must be moved out of the centered position with a current higher or lower than the null current.

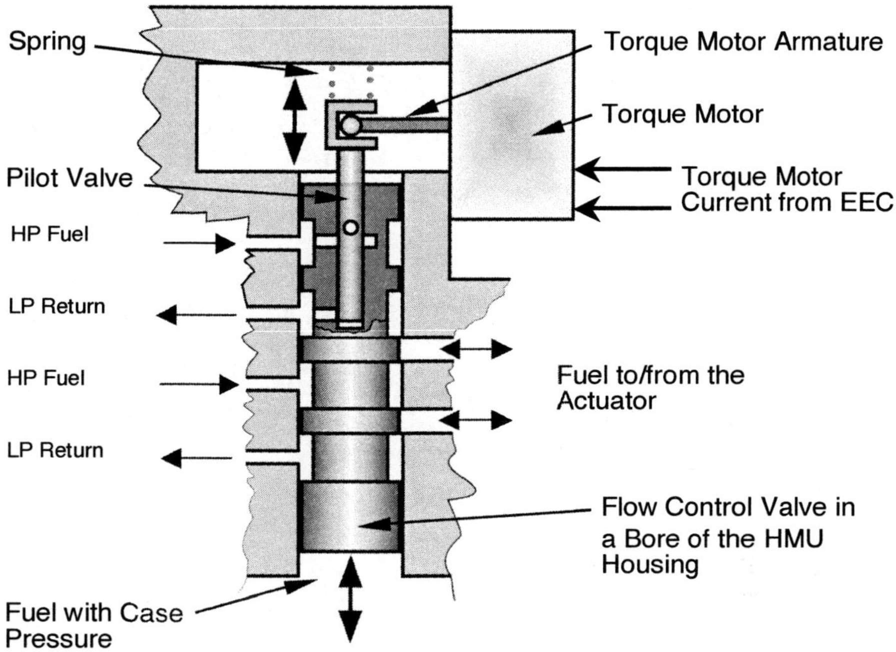

Fig. B.3 Servo valve with mechanically moved pilot valve (simplified)

The servo valve spring ensures a displacement of the servo valve in the event of an electrical failure. This displacement causes the actuator to move into its fail-safe position.

B.5 Fail-Safe Actuator Positioning

In the event of a complete loss of the EEC current to the torque motor the armature moves into its neutral position. This is not the null position of the operation. The neutral position is defined by the spring force acting on the armature. In servo valves with pilot valves a separate spring is installed. In the other servo valve types the spring force of the deformed armature is used.

When the armature has reached the neutral position the fuel pressures lead to an actuator positioning in the end position that is defined as the fail-safe position for the respective actuator.

Thus the loss of electrical control signals for an actuator from both EEC channels always leads to an actuator positioning in the fail-safe position. With an actuator in the fail-safe position the engine can be operated within the operating limits but with a limited operating range.

Appendix C
Unsuccessful Engine Starts

C.1 Types of unsuccessful Engine Starts

An unsuccessful engine start is caused by at least one of several different abnormal operating conditions. These abnormal conditions can be caused by system faults, operating errors or environmental conditions. If such an event occurs the engine is unable to start or the start sequence is interrupted to prevent the engine from being damaged. In such cases often a limit exceedance is imminent or has happened.

It is common practice to name an unsuccessful start after the behavior the engine has shown during the start sequence. The most common types of unsuccessful engine starts are:

- The Hot Start
- The Wet Start
- The Hung Start
- The Start Stall

C.1.2 The Hot Start

The EGT limit for the engine start is set on a lower temperature than the EGT limits for take-off thrust and maximum continuous thrust. The reason for this lower limit is the lower cooling airflow through the air-cooled engine parts. It is very low at the rotor speeds below idle. To prevent an overheat of the engine parts the maximum gas temperature is limited to this lower value.

During a hot start the EGT reaches or exceeds the EGT limit for engine starts. Figure C.1 shows the progression of the EGT during the start sequence. In Fig. C.2 the EGT progression during a hot start is shown. Several abnormal conditions are possible causes for a hot start:

- Too much fuel is injected during the start sequence. Fuel control system fault.

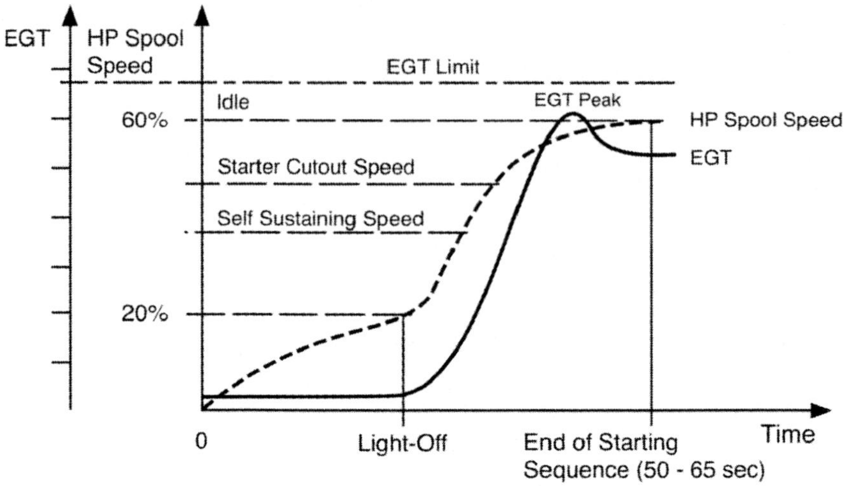

Fig. C.1 Typical EGT progress during the start sequence

- The air pressure for the starter is too low. Thus the rotor speed is too low to provide sufficient compressor airflow.
- Starter valve does not open completely, preventing proper operation of starter.
- Incomplete purging of fuel in the combustion chamber from the previous start attempt.
- Foreign Object Damage (FOD) preventing sufficient engine acceleration and airflow.
- Incorrect scheduling of variable stator vanes (VSV).

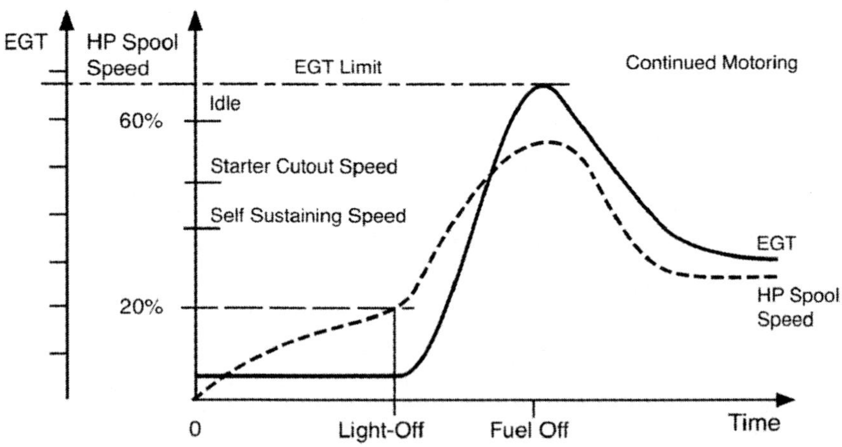

Fig. C.2 EGT progress during a hot start. The fuel was shut off before the EGT limit was exceeded

- Warm exhaust gases from another engine are ingested by the engine during start.

The consequences of a hot start depend on the peak temperature reached during the event and the time the reached temperature has persisted. In the aircraft maintenance manual a diagram helps the mechanic to determine which action is necessary after the hot start event. The possible actions range from a simple borescope inspection (small limit exceedance) to the engine removal before the next flight.

On FADEC-controlled engines the FADEC computer (EEC) prevents the exeedance of the EGT Limit during ground starts in most cases by shutting off the fuel before the limit is reached. This is possible because the EEC is able to monitor the rate of EGT increase. If this rate is too high while the EGT approaches the limit, the EEC reacts and shuts off the fuel flow. During an in-flight start the EGT exceedance is only indicated. The decision to abort the start sequence must be made by the pilot.

C.1.3 The Wet Start

When the injected fuel doesn't ignite, no light-off occurs. As a consequence the start sequence is aborted and a large amount of the injected fuel remains in the combustion chamber in liquid condition.

The reason for a wet start is a faulty ignition system. Due to the fact that an engine start on ground is normally performed with one ignition system only, the pilots realize the faulty ignition system when a wet start occurs.

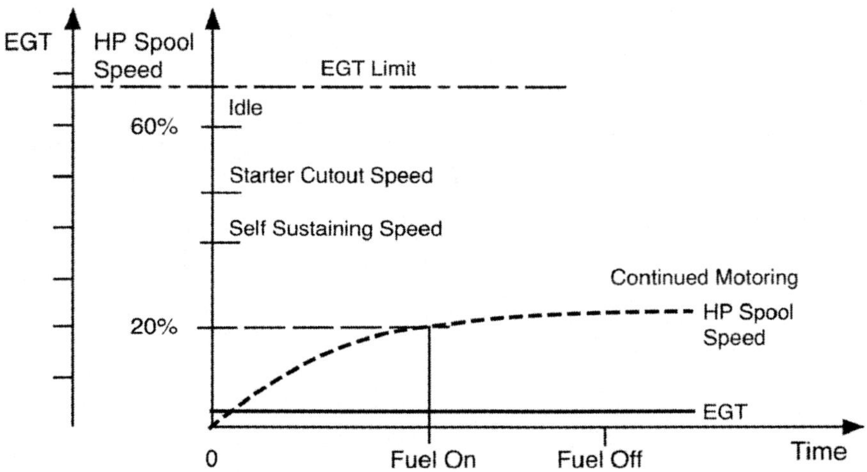

Fig. C.3 EGT and HP spool speed during a wet start

Before a new start attempt is initiated it is important that the fuel remains are removed out of the combustion chamber. This is done with a dry motoring of the engine. During a dry motoring the engine is cranked by the starter up to maximum starter speed for one or two minutes without fuel injection and ignition. The airflow through the combustion chamber blows away the liquid fuel. Without this dry motoring a hot start is very likely during the following start attempt.

FADEC engines with automatic start option react automatically to the wet start condition. The FADEC system interrupts the fuel flow while the starter is still driving the HP spool to blow out the remaining fuel. After a set time interval the FADEC system releases the fuel flow again and activates both ignition systems. The in-flight starts are always performed with both ignition systems active. This procedure prevents the loss of time necessary for the repeat of the start sequence if an ignition system fails.

C.1.4 The Hung Start

During a hung start the light-off is followed by abnormally slow acceleration and rotor speed stabilization below idle. If the normal amount of fuel is injected during a hung start it can be accompanied by an EGT limit exceedance because the fuel flow is metered for acceleration. Hung Starts can be caused by:

- Starter air pressure too low to accelerate the engine to the self sustaining speed. At the low rotor speeds the spool needs the torques of the starter and the turbine together to accelerate.

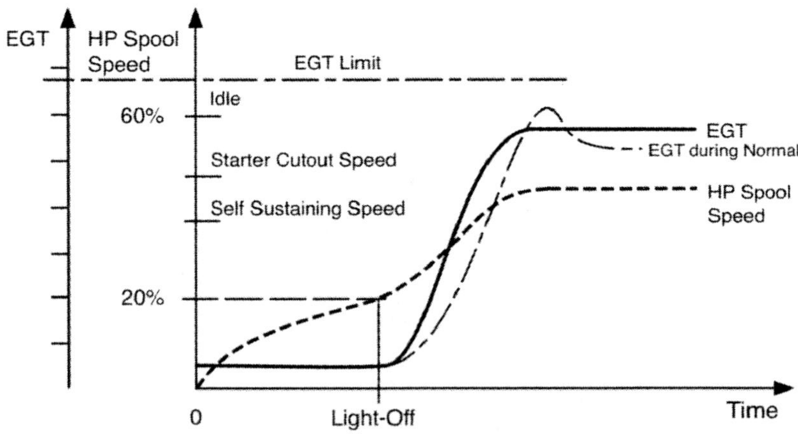

Fig. C.4 A hung start with a HP spool speed stabilized below starter cutout speed. The EGT is higher than the idle EGT and remains below the limit

- FOD to compressor. Damaged airfoils create insufficient airflow.
- The fuel flow is too low. Fuel control system fault.
- Incorrect scheduling of HP compressor IGV and variable stators. Vane angle too far to the low speed position. Results in an airflow that is too low.
- Turbine section damage. Torque from the turbine is too low due to damaged airfoils.

Before a new start attempt is initiated the cause for the hung start must be identified and corrected. An FADEC system monitors additionally to the EGT the rate of the HP spool speed increase. If this value is too low over a certain time interval the EEC detects the hung start. On some systems the EEC tries to increase the spool acceleration by increasing the fuel flow.

C.1.5 The Start Stall

At the low shaft speed above the light-off speed the surge margin of the compressor is very narrow. Small deviations from the normal hardware condition or system operation can lead to a compressor stall during the start sequence. When a stall occurs the start attempt must be aborted to prevent damage to the engine (excessive temperatures). That a start stall damages the compressor mechanically is very unlikely because the airflow is very low. Thus the forces acting on the airfoils are also low.

A start stall can be noticed by the rumbling noise, rotor speed fluctuations and EGT fluctuations. The following conditions can be causes for a start stall. They all reduce the surge margin of the compressor:

- Crosswind during engine start. Creates vortices at the air inlet that are ingested by the compressor.
- Damaged compressor airfoils.
- Tail wind condition
- Fuel flow is too high.
- Closed start bleed valves, if installed.

FADEC systems are able to detect a start stall. The EEC initiates a recovery of the compressor by interrupting the fuel flow. The starter remains on for the following automatic start attempt with a lower fuel flow. The automatic start is finally aborted if none of the start attempts is stall free.

C.2 Further Abnormal Start Conditions detectable by FADEC Systems

Additionally to the conditions of unsuccessful engine starts described above, the FADEC systems are able to detect other abnormalities that can prevent the trouble-free execution of the start sequence. A selection of these conditions are listed below:

- Low N1 detection
  - The EEC checks the N1 speed at a certain HP spool speed during the start sequence. If the N1 speed is too low a warning is indicated. Without an immediate increase in N1 speed the start sequence must be aborted.
- Sheared starter shaft detection
  - If the HP spool speed doesn't increase after a certain time interval after start valve opening, the EEC detects a sheared starter shaft and aborts the start sequence.
- Start valve fault
  - If the start valve open signal is not received by the EEC after a certain time interval after having sent the opening command to the start valve, the EEC detects a start valve fault. This detection function requires the pneumatic system air pressure in the normal range.
- Hot gas ingestion
  - If the T25 exceeds a set limit at a certain HP spool speed the ingestion of hot gas is assumed. This can be hot gas from another engine or the engine's exhaust gas during a tail wind condition. The hot gas ingestion condition is indicated on the flight deck. The operator at the controls has to expect a higher EGT peak during the start sequence.
- Thrust lever not at idle
  - Before the start sequence is initiated, it is important that the thrust lever is at the idle position. The EEC does this check at the beginning of the start sequence.

Glossary

Accessory Gearbox	The accessory gearbox drives the accessory units installed on its mounting pads. It is installed under the fan case or the core engine of the respective engine. The accessory gearbox is driven by the high pressure rotor of the engine via an angle or transfer gearbox for the redirection of the drive shaft.
Active Clearance Control	The control of the rotor blade tip clearance by air cooling of the turbine case. The cooling airflow is actively controlled by the engine control system to match the airflow to the cooling demand.
Airspray Nozzle	A fuel nozzle installed in the combustion chamber that uses the airflow around the fuel discharge for the atomization of the fuel.
Auto Flight System	The system for the automatic control of the aircraft in flight. The systems has functions for flight guidance (autopilot/flight director, autothrottle or autothrust) and for flight management.
Auto Start	Short form for automatic start procedure. During an Auto Start the complete start sequence is controlled by the FADEC system.
Autothrottle System	System for the automatic control of the engine power setting. The flight management computer controls the throttle via a servo system.

Autothrust System	System for the automatic control of the engine power setting. The flight management computer controls the engine thrust setting by sending the command data via the data bus connections to the FADEC computers of the engines. The throttles have no servo system and remain in the manual set position while the Auto Thrust System operates. This system design is used on Airbus aircraft.
Basic Engine	The basic engine consists of the main components of an engine. These are the compressors, the combustor, the turbines and the accessory gearbox.
Bleed Air	Air extracted through ports in the compressor case or the combustion case out of the main gas flow. Bleed air is used for the cooling of engine components and for the supply of the aircraft pneumatic system.
Borescope Inspection	A borescope is an endoscope for technical purposes. It operates with fibre optics or video technology. An engine has several case openings (borescope ports, plugged during engine operation) for the insertion of a borescope. During a borescope inspection the gas path of the engine is inspected for damages with a borescope through the borescope ports. No disassembly of the cases is necessary.
Compressor Control Systems	Group of systems that are necessary for the surge-free operation of a compressor.
Compressor Map	A diagramm that shows important properties of a compressor. On the y-axis the pressure ratio is shown. The x-axis shows the mass flow of the compressor. The operating points of different rotor speeds are connected by the operating line. Above the operating line the surge line is shown. The distance between the operating line and the surge line represents the surge margin of the compressor.

Corrected Value	When a value, which depends on atmospheric conditions, is measured under atmospheric conditions other than standard atmospheric conditions, it can be converted to the value that would be measured under standard atmospheric conditions. This converted value is called a corrected value.
Cowl Ventilation	The ventilation airflow between the cowling and the engine cases. It cools the accessory components and prevents the accumulation of flammable vapors under the cowlings if fluids are leaking out of a component or line.
Data Bus	A connection between two or more computers for the exchange of data. Important for the operation of a data bus is the data format used. It is given by the respective specification. In widespread use is the single direction data bus according to the ARINC 429 specification. For a two-way data communication two buses of this type are necessary. The more sophisticated data bus design is built according to the ARINC 629 specification. This is a bidirectional data bus first used in the Boeing 777. On the A380 and the Boeing 787 the ARINC 664 AFDX data bus will be used. It operates like the Ethernet network bus.
Demountable Engine	An engine ready for the installation to the airframe. All components necessary for the structural and the system interfaces are installed on the engine.
Domestic Object Damage (DOD)	Damage to engine parts (mainly in the gas path) caused by separated engine parts (domestic objects).
Dual Annular Combustor	An annular combustion chamber with dual domes separated radially. It operates with a staged fuel supply into the two dome regions. This system is used by CFM for the reduction of NOx emissions.
EGT Margin	The temperature difference between the actual EGT and the EGT limit at an ambient air temperature that equals the flat rate temperature.

Electronic Centralized Aircraft Monitor (ECAM)	The monitoring and indication system for system data used on Airbus aircraft. The monitoring function of the system reduces the workload of the flight crew for system monitoring significantly.
Electronic Engine Control (EEC)	The central digital computer of a FADEC system. It is also called Electronic Control Unit (ECU) on engines from GEAE and CFM. The EEC houses two identical computers for redundancy. These are designated Channel A and Channel B.
Electronic Indication and Crew Alerting System (EICAS)	The monitoring and indication system for system data used on Boeing aircraft.
Engine Condition Monitoring (ECM)	A procedure for the monitoring of the condition and efficiency of engines in operation. The information is acquired by the analysis of engine data measured in set intervals.
Engine Motoring	Turning of the engine rotor by the starter only. Without injection of fuel into the combustion chamber it is called dry motoring; wet motoring with fuel injection. Instead of motoring also the term cranking is used.
Engine Pressure Ratio (EPR)	The total pressure ratio across the engine. For its calculation the total pressure at the inlet of the primary nozzle is divided by the total pressure of the fan inlet. With station numbers: EPR=P5/P2
Engine Run-Up	Engine ground run for maintenance purposes.
FADEC System	An engine control system in which the control process is performed exclusively by a digital computer. No mechanical backup for the digital computer is provided and no other component than the computer is able to achieve control tasks. FADEC is an acronym for Full Authority Digital Engine Control.
Fail-Safe Position	The position an actuator automatically moves to when its control circuit is interrupted due to a fault in the electrical or hydraulical part of the system.

Fan Air	The air of the secondary airflow of a turbofan engine behind the fan stage.
Fan Frame	Designation for the compressor intermediate case between LPC and HPC in engines of GE Aviation and CFM.
Foreign Object Damage (FOD)	Damage to engine parts (mainly in the gas path) caused by objects not belonging to the engine (foreign objects). In most cases these objects are ingested by the engine from the ground.
Flat Rate Temperature	Ambient air temperature below the corner point of the flat rate thrust curve.
Fuel Control Unit	The hydromechanical fuel control device. In most cases it controls the high pressure rotor speed.
Fuel Metering Unit	The fuel metering device of a FADEC system. Its function comprises fuel metering only.
Fuel Spray Nozzle	A fuel nozzle installed in the combustion chamber. The degree of atomization depends on the fuel pressure at the fuel discharge.
Full Flow Lubrication System	A lubrication system without any pressure control device. The oil flow and the resulting oil pressure depend on pump speed and oil viscosity. Thus the oil pressure changes with changing pump speeds. Full flow systems require smaller pumps than constant pressure systems.
Gas Path Station	A reference plane in the gas path of the engine. Gas path stations are established at inlets or outlets of components like compressor, combustion chamber or turbine representing the beginning or end of working cycle sub-processes.
High Pressure Spool	Assembly made of the high pressure compressor and the high pressure turbine. Also known as high pressure rotor.
Hot Start	An engine start during which the EGT reaches or exceeds its limit. Normally an unsuccessful start because the start sequence is interrupted when the EGT reaches the limit.

Hung Start	An unsuccessful engine start attempt during which the acceleration ceases too early and the spool or shaft speed stabilizes below idle.
Hydromechanical Unit	Fuel metering device in a fuel distribution system of a FADEC-controlled engine. An HMU contains additionally to the fuel metering components the servo valves for the actuator control.
Integrated Drive Generator	An AC generator with integrated constant speed drive gear.
Intermediate Case	The engine case between the two compressors of a twin-spool engine. It is a load carrying structure because the front shaft bearings are mounted to the intermediate case.
Internal Air System	Airflow system inside the basic engine. It guides the air for component cooling and sealing through internal passages.
Linear Variable Differential Transducer (LVDT)	An LVDT is a type of electrical transformer used for measuring linear displacement.
Lubrication Unit	An accessory unit that contains the pressure pump and all scavenge pumps of the lubrication system.
Main Engine Control	A hydromechanical control unit for fuel control and the control of the subsystems like VBV, VSV and active clearance control. The term MEC is used by GEAE and CFM.
Operable Engine	A fully assembled engine with all engine related accessory units installed. In this state the engine can be operated in the engine test cell.
Parasitic Airflow	The airflow extracted from the engine gas flow for cooling and sealing purposes within the engine.
Power Plant	An aircraft engine with installed nacelle components.
Power Mangement	The calculation, which determines the nominal value of the control parameter from several input values like thrust lever angle and atmospheric conditions.

Pressure Balancing	A method for the reduction of the axial loads acting on the location bearings. Pressure differences across rotor components generate the opposing forces.
Pressure Transducer	Device that converts pressure into an electric signal. This electric signal can be an analog or digital signal.
QEC Kit	Kit of components, which are required for the installation on the engine to prepare the engine for the installation to the airframe.
Rotary Variable Differential Transducer (RVDT)	A rotary variable differential transformer is a type of electrical transformer used for measuring angular displacement
Rotor System	Assembly consisting of compressor and turbine rotor.
Rotor Shaft	Part of a rotor system.
Self Sustaining Speed	Speed of the HP spool of an engine. Above this speed the engine can accelerate without starter assistance.
Single Annular Combustor	Annular combustor with a single dome ring.
Stability Bleed Valves	Bleed valves of a compressor that ensure a surge-free operation of the compressor.
Start Bleed Valve	Stability bleed valve that is opened during the start sequence only.
Surge Margin	Area between the operating line and the surge line in the compressor map of a compressor.
TAPS Combustor	Single dome combustor design of GE Aviation with improved swirling and mixing of fuel and air. This optimizes the flame zone and its temperatures for the reduction of NOx emissions.
Thermistor	Electric component whose resistance changes significantly with its temperature.

Thermocouple	Electric component made of two different metals. Develops a voltage depending on the temperature of the component.
Thrust Lever Angle	Angular displacement of the thrust lever.
Thrust Resolver Angle	The angle sensed by the thrust lever resolver.
Thrust Specific Fuel Consumption	Absolute fuel consumption divided by the thrust related to this fuel consumption.
Torque Motor	DC motor with limited deflection of its armature.
Variable Bleed Valves	Bleed valves behind the LPC of an engine. They allow an operation of the LPC at low rotor speeds without surge.
Variable Stator Vanes	The turnable stator vanes of a compressor.
Wet Start	When no light-off occurs after the release of the fuel flow during a start attempt, this start attempt is called a wet start. A wet start results in some unburnt fuel inside the combustion chamber.

Bibliography

Actel Corporation (2005), Application Note: Developing AFDX Solutions, Actel Corporation, Mountain View, Calif. USA

Airbus Industrie, 10th Performance and Operations Conference (September 1989), Airbus Industries, Blagnac Cedex, France

Boeing Commercial Aircraft, Aircraft Maintenance Manual B737-300, Boeing Commercial Airplanes, Renton, Wash., USA

Boeing Commercial Aircraft, Aircraft Maintenance Manual B777, Boeing Commercial Airplanes, Renton, Wash., USA

CFM International(2000), CFM56-7B Engine Systems, CFM International, Melun-Montereau Cedex, France

CFM International(2001), CFM56-7B Fault Detection & Annunciation, CFM International, Melun-Montereau Cedex, France

EASA, Certification Specifications for Large Aeroplanes CS-25, Book 1 Amendment 2 (2. October 2006)

European Organisation for Civil Aviation Equipment (1992), Doc. ED-12B: Software Considerations in Airborne Systems and Equipment Certification, www.eurocae.org

Federal Aviation Administration, Joint Aircraft System/Component Code Table and Definitions (February 2002)

Gunston, Bill (1995): The Development of Jet and Turbine Aero Engines, Patrick Stephens Limited, Sparkford, Somerset, UK

Lufthansa Technical Training (2005), Training Manual, Fundamentals, M14 Propulsion

Lufthansa Technical Training (2001), Training Manual, PW4000 Basic Engine

Lufthansa Technical Training (2001), Training Manual, PW4000 Engine Systems

Lufthansa Technical Training (2000), Training Manual, CFM56-5A, -5B, -5C Thrust Reverser

Lufthansa Technical Training, Training Manual (2002), WD-2-FB-Fan Reverser CF6-80C2/-50

Lufthansa Technical Training (2000), Training Manual, V2500-A5 Thrust Reverser

Lufthansa Technical Training (2001), Training Manual, CFM56-3 Location Training Manual

Lufthansa Technical Training (2001), Training Manual, CFM56-3 Engine Systems

Lufthansa Technical Training (2001), Training Manual, CFM56-5A Basic Engine

Lufthansa Technical Training (2001), Training Manual, CFM56-5A Location Training Manual

Lufthansa Technical Training (2002), Training Manual, CFM56-5C Location Training Manual

Lufthansa Technical Training (2003), Training Manual, CFM56-7B Location Training Manual

Lufthansa Technical Training (2003), Training Manual, CFM56-7B Basic Engine

Lufthansa Technical Training (2003), Training Manual, CFM56-7B Engine Systems

Lufthansa Technical Training (2003), Training Manual, V2500-A5 Basic Engine

Lufthansa Technical Training (2001), Training Manual, V2500-A5 Location Training Manual

Lufthansa Technical Training (2003), Training Manual, Airbus A340, Rolls Royce RB211 Trent 500, Line & Base Maintenance

Lufthansa Technical Training (1995), Training Manual, Airbus A330, Rolls Royce RB211 Trent 700, Line & Base Maintenance

Lufthansa Technical Training (2001), Training Manual, Bombardier CRJ 100/200, GE CF34, Line & Base Maintenance

Exxon Mobil, Mobil Jet Oil II, Product Description, Exxon Mobil Corporation, www.exxonmobil.com

SAE Standard, AS755 (August 2004): Aircraft Propulsion System Performance Station Designation and Nomenclature

Pratt & Whitney, Service Information Report (August 2003): SOAP-One Useful Method to Monitor Oil/Engine Health, Pratt & Whitney, East Hartford, Ct., USA

Pratt & Whitney, Gas Turbine Seminar (1990). Pratt & Whitney, East Hartford, Ct., USA

Pallet E.H.J. (1996): Aircraft Instruments and Integrated Systems. Addison Wesley Longman Ltd., Harlow Essex, England, 2nd Impression

Index

G

H

I

J

L